The Leverage Principle

A Software Architect's Guide to Optimizing Development and Operations

By Mark D. Seaman

Copyright © 2015

All rights reserved.

Dedication

To Stacie, the greatest love of my life.

Table of Contents

Part 1 - Dynamics of Software Leverage

Chapter 1 - The Leverage Principle - page 9

- Software is Central to Modern Life
- Best Practices
- Sustainable Development
- Tale of Two Companies

Chapter 2 - Technical Debt - page 24

- Understanding Technical Debt
- The Principle of Balanced Development
- Multiple Types of Debt
- Recognizing Technical Debt
- Management of Technical Debt

Chapter 3 - Best Practice Leverage - page 43

- Developing Best Practices
- Promoting Software Process
- Spheres of Influence: Me, Us, Them
- Essential Practices for Leverage

Part 2 - Leverage in Development

Chapter 4 - Technology Leverage - page 58

- Technology Provides Leverage
- Technology Trends
- Technology Investment
- Road Map for Technology, Tools, Process

Chapter 5 - Architectural Leverage - page 75

- Need for Reusable Architecture
- Evolutionary Design
- The Prime Directive - Encapsulation
- Practices for Design Leverage

Chapter 6 - Code Leverage - page 96

- Tasks within Code Construction
- Test - Verify Functionality
- Fix - Repair Defects and Errors
- Extend - New Features
- Improve - Structure and Performance

Chapter 7 - Test Leverage - page 116

- Testing - Traditional or Practical
- Expected Results
- Speed Goals
- Test Cases
- Test Suites

Part 3 - Leverage in Operations

Chapter 8 - Managing Your Release Cycle - page 138

- Controlling Scope
- Managing Quality
- Shorten Release Cycles
- Continuous Delivery

Chapter 9 - Services Architecture - page 151

- Services that Scale

- Fracture Planes
- Evolutionary Architecture
- Managing Performance

Chapter 10 - Application Deployment - page 165

- Versioning
- Continuous Integration
- Configuration Management
- Provisioning Servers

Chapter 11 - Monitoring Operations - page 181

- The Metrics Mindset
- Instrumentation and Logging
- Analytics and Dashboards
- Measurements Drive Decisions
- Monitoring Tools

Part 4 - Culture of Leverage

Chapter 12 - Knowledge Management - page 199

- Leverage Understanding
- Capture - Where Ideas Go to Thrive
- Organize - Identify the Natural Connections
- Refine - Ready for the Sharks?
- Share - Time to Go Public

Chapter 13 - Teamwork - page 219

- Build a Culture of Leverage
- Attributes of Health
- Building the Desired Culture

Chapter 14 - Learning - page 231

- Learning is a Strategic Capability
- The 50 Tricks Philosophy
- Skill Mapping
- Build a Culture of Learning

Chapter 15 - Planning for Leverage - page 249

- Flexible Planning
- Functional Breakdown
- Tracking Your Progress
- Controlling Scope

Appendices

Appendix A - Build Your Own Complexity Measurement Tool - page 262

- Enumerate source code
- Measure module size
- Calculate complexity
- Estimate non-linear complexity
- Estimate module complexity

Appendix B - Testing Automation Interfaces - page 267

- Build around testing scenarios
- Stimulate the interface
- Testing with Live Data

Appendix C - Component Encapsulation - page 271

- Encapsulation & data hiding
- Extending functionality

Preface

Developing software is far more costly than it needs to be. Companies that understand and apply best practices can produce dramatically better results. Software arrives earlier to market with less cost and it also continues to work much longer.

The key difference is in the ability to leverage the understanding over multiple generations of development. This produces a repeatable and predictable development process. Building a set of best practices reverses the natural process of bit rot allowing the software to stay in operation far longer.

The Leverage Principle builds a case for establishing best practices in three important areas. We explore the most common problems that block leverage and propose solutions in the form of best practices. This book is meant to provide solid advice to people actively doing software architecture and design.

Architectural Concerns

This section delves into issues related to the services you are building. We discuss how to manage technology over time and also explore ways to create architectures that can be used over multiple generations of products. We also look at the critical practices for reducing the software decay process. There is an emphasis on testing strategies that probe the code from different directions to create an architectural resilience that enables software systems to last for many years.

Operational Concerns

Operational integrity is a vital concern for every software development entity. We will explore the release cycle and best practices associated with continuous delivery. We will look at techniques for managing and optimizing the health of the software over its entire lifetime. These chapters outline the proven tools and practices that lower operational costs.

People Concerns

People issues affect the way that software is developed. Since knowledge is the primary asset, managing learning within your team is essential to achieving any business goal. These chapters demonstrate how to manage the sharing of knowledge and planning to support higher levels of leverage within your organization.

The artwork on each Part page features one of the Six Simple machines defined by the Renaissance scientists. The lever, wheel and axle, pulley, incline plane, wedge, and screw were the earliest inventions that allowed humans to accomplish more work with the same effort. As such, they are a visual reminder of the power of tools to create greater things in every human endeavor.

Acknowledgments

This book has been in my head and heart for many years. The itch to write a book is common to many people but to actually complete it has taken the input, guidance, and encouragement of many, many people.

I want to thank my wife, Stacie Seaman, for her encouragement and numerous contributions throughout the process of writing and producing this book.

Thank you to my business partner, Eric Williams, for his support. He kept Shrinking World Solutions running smoothly, allowing me to dedicate the time I needed to write this book.

I also want to thank the colleagues that agreed to serve as reviewers. Your input was essential in identifying areas that were weak and adding insights and tools in specialized areas. Thank you Julio Garcia, Greg Brake, Jim Edwards, and Josiah Seaman.

PART 1 - Dynamics of Software Leverage

Figure 1:

Chapter 1 - The Leverage Principle

"Give me a lever long enough and a fulcrum on which to place it, and I shall move the world."

~ Archimedes

Software is Central to Modern Life

In 2011, Marc Andreessen famously stated that "Software is Eating the World". Every industry has experienced a remarkable transformation due to the growing dominance of software. Entire industries like bookstores, music stores, movie studios, newspapers, photography, and financial services, have been replaced by software-based services. Software is no longer a secondary player - it's the main show and will dictate whether your business succeeds or fails. Business now runs 24/7 and software is at the very heart of this amazing revolution.

Great software systems create business success - they can automate repetitive tasks and free up resources to continue innovating in other areas. Humans are great at solving problems and computers are great at running those solutions. With each problem that is solved, more opportunity is created to deliver greater value. The business now has a solution that can keep delivering value for an ever decreasing cost. Resources and attention can now pivot to solving the next round of business problems. Innovation spawns more innovation and forms a productive feedback loop.

The reverse is also true - inadequate software can easily destroy an otherwise healthy business. Because software is so central to every aspect of business today, a company can be crippled by software systems that don't meet the business needs.

Think about the national healthcare initiative, healthcare.gov, and its high profile flop. Thousands of hours of planning and development costing millions of dollars culminated in a failed deployment of a web application. The failure created extra scrutiny, loss of credibility, and several key leaders lost their jobs.

This drama plays out every day in every industry. Companies, governments, and other organizations that don't understand how to develop or purchase software systems that match critical business needs are doomed to struggle. The answer to this threat is to apply the leverage principle to every facet of software development and operations.

Importance of Leverage One definition of leverage is using something to maximum advantage. The goal for every software development project should be to utilize every asset - everything you've previously built, everything your team learned about the business needs, every skill and bit of expertise in a technology platform. Utilize these previous investments rather than creating them new every single time. No one can afford to rebuild infrastructure that has already been created.

The Leverage Principle - *Best practices produce higher quality software which encourages reuse.*

Software development is expensive and it is far more expensive than it needs to be. To build a software system requires solving thousands of specific engineering problems. Each of these problems require some minimum amount of effort. This effort is the essential problem cost.

But the cost of the actual solution can far exceed the essential cost. Inefficiencies in the development process introduce work that doesn't actually contribute toward the project goals. This cost must be viewed as waste. As much as 50% of the ultimate cost of any software development project could be characterized as waste. Eliminating this waste can allow you to deliver far more business value with the same resources.

Cost and Leverage There is a direct relationship between the incremental development cost (in money and time) and the amount of leverage that is achieved. The following graph shows how the cost of the next release can be contained by reusing the understanding from the existing system. High leverage levels require a lot of discipline that must be acquired over time. But even modest amounts of leverage can have a dramatic impact on the overall development cost.

You might think that it is hard to achieve high levels of leverage on each new product cycle, but the opposite is actually true. The leverage is determined by the quality of the software development process which is driven by the extent of best practice usage. There is a compounding effect that applies here - better processes enable a greater rate of improvement. Each generation will see a reduction in the cost based on the developed skill and discipline of the team. There is a double benefit here. Improving software processes produces better software in the short term, but it also reduces the cost in future development.

Teams that have already achieved a high amount of leverage will continue to improve their processes at a rapid rate. Organizations that are struggling with the fundamentals of software engineering will find it difficult to realize the leverage goals simply because they don't know how to develop software. You can expect your rate of improvement to be relatively consistent over multiple product generations. This gives you an idea of what to expect for cost improvement over time. The budget contraction for money and time is the truest measure of leverage.

Rate of Cost Improvement

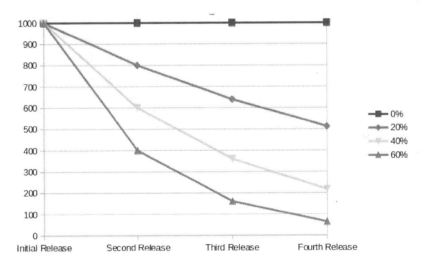

Figure 2: Rate of process improvement

What Blocks Leverage? Everyone desires and expects leverage during software development. But when we begin to look at reality, our projects often disappoint us. Why don't we get more leverage? The answer is technical debt. A new project presents us with an abundance of problems that must be solved. Each day on the project we are working against this backlog. Each problem that we work on may reveal other problems to solve. We can think about every one of these unsolved problems as a debt that is owed by the project. Eventually we work through the debt and the product is released to market.

In practice, we choose to solve some problems while deferring others. Every time we defer an issue for later, we write an IOU that must be paid some day. The best way to measure both assets and debt is hours of engineering required. How many hours will it take for you to resolve the issues on the project? How many hours would it take to build the functionality that you have right now? This gives you a solid metric that accounts for what is good and bad within your software.

Technical Debt Decreases Quality Excessive technical debt is the primary challenge to quality. Unseen problems multiply and threaten to compromise the integrity of the system. Quality is often assumed (of course we write good code!), so very few people will appreciate it unless it is missing. Quality is not automatic, it requires hard work. Groups that produce high quality products will continue to do so because they have discovered the required practices that lead them to success.

The only way to have repeatable results is to monitor and control your technical debt throughout the course of your project. Quality problems propagate - a small problem can grow into a large problem if left unresolved. Bugs reproduce and infest other parts of the system. Unclear designs create more confused code. Misunderstood product requirements result in solutions that work "correctly" but aren't usable.

Eliminate debt as you go in order to pay the lowest cost. A bug that costs you an hour now may cost you a day to fix one week from now. That same defect could easily cost you a week of time later in the project. The strategy should be pay as you go - clean up every area you visit before moving on.

Make debt a conscious decision - don't accept it automatically. There are real situations that require postponing work until later. Choose that path as a last resort and be aware that you will have to pay interest on that work in addition to the initial cost.

Many Types of Technical Debt   There are many types of technical debt. There are different sources of debt that show up in many ways. The actions for solving each type of debt are unique to the problems being addressed. The structure of this book is based on addressing the three main categories of debt:

- Development Concerns - Part 2
- Operational Concerns - Part 3
- People/Culture Concerns - Part 4

We will look at each of these areas from an architect's viewpoint and ask, "how can we apply best practices to control the level of debt that is accumulated over time"? Paying off technical debt produces the maximum opportunity for leverage. Technical debt is eliminated by having better development practices that standardize the approach to common problems. This decreases the number of unique problems that must be solved throughout the development cycle. The amount of leverage determines the cost of software. Best practices are the most direct way to control software cost.

Best Practices

Technical debt is a trap that will prevent leverage. Establishing standard practices is the most effective path to avoid debt. When best practices are implemented throughout the development process, creative energies are focused solely on solving new problems.

Most architects are familiar with the excellent work done in the book, "Design Patterns: Elements of Reusable Object-Oriented Software". It introduced us to the ideas behind creating a catalog of designs for common problems. We can extend this concept to every other aspect of software development. Design patterns represent leverage in the design space but we also need leverage when it comes to Requirements, Coding,

Testing, Skill Management, Project Planning, Operations Planning, and Monitoring Techniques.

Every aspect of development could benefit from producing canned solutions to common situations. In other words, we need Requirements Patterns, Operational Patterns, Skill Management Patterns and others. The goal should be to have a complete set of process patterns to share knowledge effectively between developers.

Creating a catalog of standard practices eliminates the need for each engineer to learn the best way to do any given task on their own. The catalog itself becomes a form of leverage allowing you to fully utilize each new trick that is learned. Imagine bringing new team members up to speed by presenting them with a catalog of the best practices they will need to do their job and jumpstart their productivity on the project.

Automatic Response to Common Problems If a conscious decision is required to accomplish every task, then the level of productivity will be fairly low. However, work accelerates dramatically when the steps to solve a problem are automatic. Remember learning how to drive? Every single action required a thoughtful analysis. If we make most development tasks automatic and unconscious then problems will be solved with very little effort.

Best practices teach us to identify the situation and then select the appropriate response. The solution is fixed and automatic for any given solution. Of course, being good at selecting and applying the patterns still takes practice. With a standard pattern, the engineer no longer needs to invent the solution, they simple apply it. This can save an enormous amount of time over the course of a large project.

Fight the pull of "Not Invented Here" (NIH). Engineers love to invent so there is a tendency to invent everything that is necessary on a project. "Use a standard toolkit"? "Not invented here"! The core flaw in this thinking is that engineering resources are being wasted on finding new solutions when existing ones work just as well. Engineering, at its core, is simply problem solving. Developing a new product requires that thousands of individual problems be solved. Successful leverage is the ability to identify these problems and implement a solution without solving every technical problem from scratch.

Role of Architect According to Wikipedia, a Software Architect is "a software expert who makes high-level design choices and dictates technical standards, including software coding standards, tools, and platforms". CNN Money adds, "Great software architects are designers and diplomats. They create innovative and valuable programs, but they also translate highly technical plans into a vision the C-suite can understand. They are a crucial link between a company's tech unit and management".

A key role of the architect is to be the primary protector of the software development investment. Leverage is central to achieving the highest level of value for a reasonable cost. Therefore, the software architect needs to ensure that the architectural design is suitable for the current release and can also be used as the basis for all future releases. This is important for small scale software projects but becomes absolutely critical as the size and scope of the projects grow.

Architects have a primary responsibility to lead the organization's development practices. Software development techniques should be constantly improving and the architect steers these efforts. True improvements have to be based on the current reality and measurement is a good starting point. A baseline will help you understand your current process and where improvements are needed.

Measure Your Current Leverage Think about your last four development projects. Estimate the budget in hours for each project. Multiply the extended size of the team by the duration of the project and convert into hours. This gives you a good snapshot of the cost of those projects. Now let's examine the leverage.

Between each successive budget, what was the percentage reduction? For example, 20% reduction would correspond to budgets of 1000, 800, 640, 512. You can see that even a modest leverage rate still produces a 50% reduction over four generations.

Now for the hard part. How similar were the problems that were solved? Very similar problems can result in an expected leverage opportunity of 90%. This is because 90% of the problems that must be solved already have a standard solution that can simply be applied to the new product.

Leverage is the result of identifying these problems and solving them in an efficient way.

Setting a Target for Leverage Each software product is tied to a domain. By exploiting the repetition of problems within a given domain and identifying the common problems and reusable components, each new generation of software will require fewer new components to be created.

On these recent projects, how much leverage was missed? How many problems did you solve for a second time without any real benefit over the previous solution? Switching technologies or programming languages is the type of large change that destroys leverage. If you could time travel, would you make different decisions?

When solving each problem, think about the problem itself rather than the solution. How much of the problem is actually different between the generations? Does this correspond with the level of leverage that you were able to achieve? If not, there is a gap that is actually an opportunity. You can speed up development significantly by exploiting more leverage opportunities. This is where the role of architect is critical to your company's future success.

Reliable Knowledge Transfer Software is about understanding how to solve certain kinds of problems. There are many different kinds of knowledge that must be managed to bring software to life. Between each generation of software development there is the potential to lose that knowledge. The team then has to reacquire lost knowledge in order to complete the next product. It is expensive to learn information the first time and it is just as expensive to relearn it a second and third time.

Leverage can be viewed as the high-fidelity transfer of understanding from one product generation to the next. If we can transfer 90% of the understanding from one product to another then only 10% of the problems must be solved. We would expect this level of leverage to result in a ten times reduction in budget.

Leverage is about knowledge transfer. When knowledge is lost, leverage is lost too. Changing technologies, product domains, organizational structure, and knowledgeable individuals can destroy your leverage potential. Sacrificing leverage can easily lead to a bloated budget and a failed product.

Sustainable Development

If leverage is using something to maximum advantage, our goal is to create a repeatable software development process. An accidental success on one project is nice but we want predictable success. This requires a deeper understanding of software process. Each generation of development should not only produce solid products but also advance our development capability.

Winning organizations strike a balance between meeting the immediate needs and solving long-term problems that will meet the future needs of the business. This is accomplished by building effective software development capability. Each product that is produced should make it easier to produce similar products for much lower cost. Leverage of the software development process is the the surest path to sustainable development. Core software capabilities support the business goals, leading to ongoing success.

Productive Improvement Cycle Improvements have a compounding effect. Each problem that is solved reveals new problems that have been hidden, but the new problems are both smaller and easier to solve with the new optimized system. Each iteration of improvement not only solves a real problem, but makes it easier to solve future problems as well.

This creates a productive feedback loop. The rate of the improvements increases with a compounded benefit. The corollary is also true - a system in decay is caught within a destructive feedback loop. The worse things get, the more rapidly they will go bad.

We can harness this dynamic to create improvements that are easy to justify based on short-term results. The key requirement is to produce results quickly. Massive efforts at process improvement frequently fail because the decision makers don't have a value for process - however they do value better product results. The clever architect will focus attention on the product results and use that to justify the process improvements.

Use Metrics to Define Organizational Norms Many organizations may be content with their current results but they have no objective way of measuring the productivity or quality. Establishing a few simple measurements can have a profound impact on how your team behaves. Begin to count a few things that are essential to your success and use this information to gain more support for your improvements.

If you show measurable progress from your improvements you can build trust with stakeholders and make more changes. This is also helpful for identifying neglected problem areas. For example, assume you can show that a particular module is ten times more complex and producing more bugs than any other module. You then have a clear mandate that this module needs to be replaced.

Build tools to examine your systems. Pay attention to the pain - it is indicating a problem that needs attention. Every disagreement or misunderstanding that happens during the course of a project is pointing you to an unsolved problem. Solve the immediate problem and create a standard solution at the same time. This solves a problem now and solves it for the future too. Over time, an organization can build a very valuable inventory of common solutions.

Seek Leverage as a Primary Goal As improvements are made daily to the features, tests, and structure, the software will continue to get better and better. Fixing an existing system will prevent having to replace it. Constantly monitor the big picture to manage both quality and functionality throughout development.

A little cleaner is a little better. You don't need to leap the building in a single bound but you do need to make a constant series of improvements to the software over time.

The economics of leverage are simple:

- High quality means low cost
- Low quality means short life

- Software reflects the team that built it
- Building best practices of the team extends software life

A Culture of Success Great software is produced by great teams. An effective team can be relied upon to create amazing software every time. A strong development team can achieve three to four times the results of a mediocre team. Building a winning team isn't easy because there are always many challenges that threaten to undermine the effectiveness of the team. Some of these challenges have become so entrenched in the organization that they aren't even recognized anymore.

The chapters in Part 4 are dedicated to exploring the people-related issues that affect leverage. Without addressing these topics you won't be able to realize the full benefits of leverage. Your success at development and operations is directly dependent on your ability to build a healthy culture. How your people think and behave will either support or undermine your business goals and technical goals. Sustainable development requires building a great place to work at the same time great products are being built.

Tale of Two Companies

To illustrate the Leverage Principle, let us consider two different fictional companies and how they might apply leverage. We will start with a small scale application and then consider how these same dynamics would apply to much larger software development efforts.

George and Linda decide to create a company, Seabreeze Travel, that will manage bookings on cruises. They decide to build a simple enterprise application which will be hosted in the cloud. They both have a lot of web development experience and are enthused about doing it right this time around. They want to build a platform that can be used long-term to grow their business. Based on their previous experience, they think that an application will require around 1000 hours of work to build.

Mary and James also decide to start a cruise booking company, called Foulwind Adventures. They attended a seminar on Agile software methods recently and they are very enthusiastic about that approach to development. They believe that being agile means not writing any documentation or doing any design up front. They select a sprint cycle of one week and load it up with lots of features. Because they have chosen to implement so much functionality they probably won't have time to write tests as part of each sprint but they aren't concerned about that because they plan to hire several testers just prior to the product release.

Startup Decisions At Seabreeze, Linda begins the work of project planning. She divides the 1000 hour budget into ten 100-hour sprints. She and George will work together on a one-week cycle. They expect to be done in around three months. They compose the milestones so that one significant feature is budgeted for each iteration. They will build out each feature fully before starting on the next.

James is a wizard at coding so he jumps right in to the first chunk of code. At Foulwind, they hope to be done in a few weeks. They already have made commitments to others that the code will be released within two months. They are feeling very excited about their impending success. Over the next four weeks they pound away on features and are able to produce about twice the functionality that Seabreeze is able to accomplish.

After about a month, Seabreeze has about one-third of the functionality created. Each feature is well-tested with design patterns and strong interfaces. In fact, Linda has created an automation framework and language to remotely run reservation scenarios from test scripts. They are excited about how this design will adjust to multiple front-ends someday. They currently have no outstanding defects.

Back at Foulwind, Mary has implemented a ton of new features. Of course, she hasn't had time to fix all of the quirks but they have saved so much time that they are confident that it will be easy to fix whatever needs to be fixed later. Mary and James have started working independently to avoid the aggravation of code conflicts - these multiple code streams can be reconciled later. It is starting to bug both of them that they don't agree on the number of spaces and variable naming conventions so they have fallen into the habit of reformatting the code to assert their favorite style.

At three months both companies release code. Foulwind was aiming for a two- month cycle but just couldn't pull it off. There were lots of strange behaviors that required a lot of debugging to resolve. They hope things are better now, but you never really know. Both companies end up with fairly good releases and customers begin to flock to the systems.

Second Release After a month of operations both companies are missing key business needs. They were both missing two critical features and are having problems with scaling to the unexpected user demand. Linda looks over the new functionality that is needed and determines that they already have 80% of what is needed for the second release. Seabreeze immediately starts work on the new release by working on two weekly sprints. Two weeks and 200 hours later they have a new release for customers.

At Foulwind there are lots of defects, which is OK because new software is always buggy. Customers are complaining about performance but the architecture doesn't really support scaling. In fact, a new set of tools and database structure may be needed to address the performance needs. All of these factors together make it difficult to reuse the existing application code. But both James and Mary believe that reuse is important so they push forward.

By the time they get to the second release they need to spend an additional budget of 300 hours to maintain the existing system. The new functionality required about 700 hours to build since there was only an opportunity to leverage about 30% of the previous design work. The concept of the sprint was abandoned because it is easier to just work on the most urgent issues which seem to vary every day. Foulwind's second release required 1000 hours of effort and three months.

The third release follows the same trend. Seabreeze is able to spend another 200 hours over two weeks to update their software. They are already thinking in terms of continuous delivery. Why not just release every two weeks?

Foulwind is starting to believe that they may have the wrong tools. They are still having performance problems and decide to switch web frameworks. They decide to switch from ASP and C#, to Ruby on Rails. They both like new technology so they believe this will be a good change of pace. The third release is a complete reset and requires 1000 hours over three months.

As they compare the results of the second and third releases, they realize the cost is the same. Both releases cost 1000 hours so they conclude there is no benefit from leverage. "Reuse is a myth! Let's rewrite from scratch each time". While this approach does have some merits, it is based on a fallacy.

Comparison of Results   The following table shows the comparison for the results produced by the two companies. The two companies start with the same level of performance. But over time the productivity diverges dramatically. Within three product cycles there is a cost difference of five times for each incremental release.

Economics of Leverage

Release	Seabreeze Cost	Foulwind Cost
#1	1000 in 3 months	1000 in 3 months
#2	200 in 0.5 months	1000 in 3 months
#3	200 in 0.5 months	1000 in 3 months
Total	1400 in 4 months	3000 in 9 months
Leverage	80% leverage	0% leverage

This illustrates the Leverage Principle. Trying to leverage a system with high technical debt is impossible. If the software is that bad it's much simpler to replace the system entirely.

Best Practice #1 - *Make leverage your primary goal.*

Problem   The key to controlling cost is fully utilizing knowledge that you already possess. The understanding of the customer needs, product domain, technology, tools, architecture, common problems, test strategies, and release process are vital to the next software project. Learning and mastering all of these areas requires a large amount of effort. A product that fully leverages that knowledge requires far less time and money and results in a very high quality product.

Yet, most projects don't leverage any significant amount in each project cycle. Knowledge is lost or discarded and must be reacquired at a high cost to the project. The fundamental problem is that leverage is an afterthought and not really built into the primary project goals.

Signs of poor leverage include:

- Development work is often repeated
- Knowledge of how to build the system is frequently lost
- Legacy code becomes very brittle over time
- Unable to reuse software because of rigidity

Solution The solution to this waste is to build leverage into your core project plan. Define all other project goals to optimize the amount of knowledge that can be applied. Avoid the wasted effort that is required to relearn things that you already know.

Make a list of the knowledge assets that you already possess and the new ones that are required to complete the project. Build your project planning to account for learning the missing essential knowledge. Learning is a significant investment so work to fill the learning gap early in the project cycle and then create experiments to validate your learning. Verify key assumptions that could threaten leverage.

Next Steps

- Identify the critical types of technical debt
- Build an inventory of best practices
- Plan necessary process improvements
- Build multi-generation product plans with optimal leverage

Chapter 2 - Technical Debt

"We wished to operate, as much as possible, on a pay-as-you-go basis, that our growth be financed by our earnings and not by debt."

~ *Bill Hewlett*

Understanding Technical Debt

Technical debt is a lot like like financial debt. Tech giants Bill Hewlett and David Packard decided on a very controversial path by rejecting using debt to grow their business. This counter-cultural stance contributed significantly to the financial stability and early success of Hewlett-Packard. As architects, we must also make a similar stand to protect our systems from the Siren's call of easy results at the expense of technical debt and the resulting rapid decay.

Before we can figure out how to decrease the technical debt on a project we have to find it. Martin Fowler's classic book, "Refactoring: Improving the Design of Existing Code", provides us with excellent advice on how to keep our code clean.

His analogy is when cleaning house we use our nose to detect bad smells that need to be addressed. We can learn to detect bad things within our code by learning to identify the bad smells. This is certainly true of old code but it can also be generalized to the broader scope of project debt. Some projects may have very little code debt and massive planning debt. Learning to recognize the full range of bad smells that emanate from a project is a valuable skill.

Problem Solving Dynamics Debt starts with a problem. If we assume that engineers can often solve a single problem in an hour, then an engineer can solve roughly 2,000 problems every year. But what if the project calls for you to solve 4,000 problems every year?

Problems happen in every domain and they all compete for a limited bank of mindshare:

- Definition of the product
- Implementation of new features
- Writing tests
- Supervising the test execution
- Sending out early marketing units
- Attending mandatory corporate training sessions
- Writing TPS reports
- Reporting scrum status
- Preparing for the visit of the VP
- Addressing customer issues for the existing product
- Investigating a new tool for continuous delivery
- Making sure that everyone is able to use the new Integrated Development Environment

The list is infinite but each engineer only has one brain. Which 2,000 problems will you solve over the next year? You will choose wisely by selecting the problems that bring you the biggest return on investment. Do the easy stuff - but continue to do some of the hard things as well. You can't possibly do everything so always look for the highest ROI.

It is Monday morning and you have just updated your action list. You have 30 items that you really need to get done today. Is this realistic? You can tell yourself that you will really try hard today and get it all done but you know it is impossible.

Make sure that the eight to ten things that you actually address are the most critical. Everything else will have to wait or go to someone else. Remember, when something should really get done and doesn't make the cut, you are creating a project IOU for the item and your debt just increased. This might be a defect in the code or a design alternative to investigate or a communication to be sent to upper management. There are many actions that must be taken and every deferred action becomes a debt to the project.

Someday you must give an account. Try to defer the tasks that don't produce a critical value for the project. Neglect some of the busywork that is promoted within every organization in favor of tasks that get you closer to business success.

Work Left Undone I once worked on a project where a massive amount of code had been written by a brilliant individual in a top-of-mind orgy of complexity. There was a single function that was 3000 lines long. I printed it out so that I could review and understand the structure. The nesting of the indent levels was so great that the code would wrap horizontally off the end of the line. It required multiple lines of output to show one line of text. That day I learned what massive technical debt looks like.

The tragedy of this situation was that the functionality of the code was indeed brilliant - the business value of the system was very high. But the coding debt was large enough to bring the net value of the system into question.

What is the debt? If debt occurs when you write the IOU, then its value can be quantified. What will it take (in hours) to fix the problem? This will let you easily measure any type of debt in your system.

In the project I just referenced, I spent about 40 hours refactoring the massive knot in the code. The system represented about 2,000 hours of development. We also had some other problems that would account for a total of 200 hours of additional repair work. After 200 hours of investment we were able to capture 2,000 hours of benefit. Without that payoff the system would have been doomed to early replacement. This looks like an ROI of ten times for the debt payoff effort and is fairly typical of projects.

```
Net Worth = Business Value - Effort to eliminate debt
```

Another scenario that is very common is a system that has accumulated a debt so massive that it far exceeds the value of the system itself. The benefit (Business Value) of the system can be measured by looking at the replacement cost.

Estimating System Value An enterprise level web application of reasonably small scope can be built in about 1,000 hours of engineering by a competent team. Many systems exist that required 10,000+ hours to build over several years. Although the cost to create this system was actually 10,000 hours, the replacement cost may be much lower if the system is rebuilt from scratch. In this case, we are replacing the code but there is a large amount of leverage from the first version in the form of same programming language, existing team, established development processes, and knowledge of the business needs. Assuming a replacement cost of only 1,000 hours, the value of the system to the business is 1,000 hours of development time.

Notice that the actual business value is far less than the actual cost of the existing system. This is because the entire system can be replaced for a fraction of the original cost. The existing system may require another 2,000 hours of work to get it in good shape so the technical debt can be calculated at 2,000 hours. The business would be better off replacing the system than fixing the debt. A system must have a great deal of value *and* low debt in order to justify the cost of maintenance and extension.

Businesses will frequently over-value the system because of the high sunk cost to get to the initial product. They will continue to invest in legacy software that has exceeded its useful life for fear of not being able to successfully create a replacement. We stick with the evil we know rather than starting with a fresh approach. This is where leverage comes in - by utilizing all of the understanding that has been gained with the first version we can build the new system for a fraction of the original cost. We leverage our understanding into a fresh implementation that doesn't have any technical debt.

The Principle of Balanced Development

Software development requires that we balance our goals between different dimensions. If all of our activity is concentrated in a single area, such as coding or testing, then our project becomes lopsided. Our work must be distributed properly to avoid technical debt.

The first dimensions that we must balance are investing in adding functionality to the product *or* investing in building quality. Neither of these are optional - we can't have a product that is missing key features but is well polished and refined. It is also isn't acceptable to add lots of features to a product with a flawed foundation. A high quality product will have a balanced investment of 50% for functionality and 50% for quality. Both features and the underlying structure should get equal attention every day.

It isn't possible to do these in sequence, they must be done in parallel throughout the development of the product. Improve the structure to make it easier to add features, then fix the defects before optimizing the structure. Repeating this cycle throughout the project leads to maximum leverage. If your development practices follow this model you will never lose quality in pursuit of new functionality *and* you won't be stalled out by endless refinement.

Below the Water Line Quality is often unseen but it is still critical to success. Some work will be highly appreciated by your customers, clients, and bosses, but much of your work can't be appreciated because it is hidden. Junior engineers often make the mistake of working only on the visible parts because that is what gets rewarded and praised. This is a serious mistake that puts the project in danger of creating a bug farm. If we only do what our boss can appreciate we produce something that no one will appreciate at all.

Managers are highly unlikely to ask engineers to refactor a piece of code so that it's cleaner. Often it's a fight to be allowed to invest in refactoring. If the software architect doesn't work on areas that are invisible and under- appreciated we have to question whether they are functioning as an architect. Certain things must be done simply because it is the right thing to do. If the architect doesn't care whether the architecture serves the long-term business interests of the company, who will? 50% of the day for an architect should be spent thinking about the part of the system that no one sees because that hidden foundation protects the part of the system that everyone else sees and values.

As Robert Martin at Object Mentor (a.k.a. Uncle Bob) taught us, even boy scouts have a motto to leave the campsite in better condition than they found it. Physicians take a vow to "Do no harm". Perhaps it's time for software developers to commit to the same code of conduct?

Now let's combine these two areas of balance to discuss the daily tasks.

Balanced Development By arranging the tasks on the following axis we can define a set of activities that must be balanced against each other. The typical day should start off with testing. What changed while you weren't looking? Fix everything you find that you don't like. Remember, fixing a bug now will keep it from breeding and save you from fixing 100 bugs later.

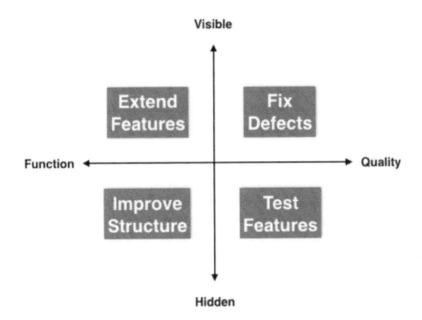

Figure 3: Development Task Types

Now figure out which new feature is the highest priority. Extend the code to implement this feature. Set yourself a time limit and be satisfied with the level of functionality that you're able to add during the limited time. This will prevent building anything that you don't *really* need.

Now focus on improving the structure. Find any hint of duplication in the area that you were just working in. Try to reduce the number of lines of code in half. Imagine that your biggest critic is about to examine the code - now clean up anything that would be an embarrassment.

Repeat this cycle until you are done. This pattern will work well at several different levels of granularity. Imagine a daily or weekly cycle. It also works well at short intervals. I often work for closure every hour because this keeps the amount of work in process to an absolute minimum and allows me to deploy new features within an hour.

Experiment with this cyclical work. Find which types of work can be done optimally with each rhythm and use what works for you. I find that no cycle is too short. The important thing is to allow the full range of actions to be fully balanced.

The other benefit is that a steady rhythm of predictable cycles will do wonders for building trust. Over a few cycles people will begin to rely on the rate at which your team can achieve high-quality delivery of functionality. This is the best recipe for establishing or restoring trust within your organization.

If you let others set the complete agenda for your day you will spend all of your time extending features and a little bit of time fixing defects. Ultimately, the success of your projects will depend far more on what you do below the water-line.

Practical Reality Let's be honest. We often don't have the luxury of solving all of the problems that exist. Often we must concentrate on the critical issues and neglect secondary issues. By definition, this creates debt and debt blocks leverage. But you don't need to let the debt accumulate unchecked. Find ways to measure the most critical forms of debt within your system. Use these numbers to create an accounting. Track a score card by counting things like unimplemented features, unresolved defects, excessive complexity, unwritten tests, and low test coverage.

Make sure that these items get just as much mindshare as the new features that were implemented in the last scrum cycle. Studies have shown that people spend more when they pay with a credit card than with cash because it is so easy and they don't "feel" the money being spent. A similar dynamic affects projects - they will always choose debt unless you make it painfully clear that there is a balance that will come due and must be paid.

We make trade-offs in every project - there is a time to ignore certain things in favor of others. But we should do this consciously - by deliberate choice. Don't just accept technical debt automatically without regard to its eventual impact on the product - create a plan for when you are going to pay it off.

Debt Can be Useful Debt itself can be a form of leverage. You may choose to implement a bad architecture that you know must be replaced eventually. Debt makes sense when it leads you to a higher goal. For instance, neglect writing a group of tests to get your coverage up. Or create a disposable prototype to investigate a feasibility issue. But don't expect these things to be in the final product - have a plan to pay off the shortcuts.

The acceptable level of risk changes throughout the project. At the beginning of a project the risk is sky high. Very little is known and lots of unknown challenges block the way to release. A well-run project lowers the risk constantly throughout the development cycle. Key questions are answered by validating the technology, proving market feasibility, and building prototypes.

A poorly run project will leave many questions unanswered and allows the risk of unknowns to remain high or even increase throughout the project. This type of risk often causes projects to miss delivery dates and goals.

Mistakes made and problems deferred can create a risk that is different in nature to the unknown risk. Unknown risk can diminish throughout the project while the risk due to technical debt can build - think about this as the knowable risk. Measure both types of technical debt and translate that knowledge into plans to pay off the debt. You can know and reduce the amount of technical debt that you carry - it should never be a surprise at the end of the project.

All types of technical debt represent a risk to your project. There is a low to high probability that a bad thing might occur and the severity of the consequence can range from serious to minor. The expected value of the risk is the product of severity and likelihood. Try to pay off the risks with the highest expected value first. Ultimately, you are playing a probability game - the smaller your exposure, the better.

Wishful thinking is deadly. You can gamble with the project and assume that you can consistently take on risks without being affected by it but someday the debt for all of the trade-offs that have been made will come due. This may occur at a time in the project when you can least afford it so be careful with how you handle debt.

Multiple Types of Debt

Debt comes in many forms and engineers experience each of these to varying degrees on every project. Identifying the debt is an important first step in managing it. Once you can see it you can begin to control it. It is often useful to trade one type of debt for another to spread out the risk. Figure out where your project is the most vulnerable and seek to mitigate that area of risk. For example, if you determine that lack of testing poses the most serious threat then refuse to increase your testing debt any further.

Each type of debt has an antidote - you can develop tools that will let you objectively measure each type of debt within your project. This will require some effort to set up but it'll have a huge impact on every aspect of the project.

Development Debt The first major category of debt is about the process of developing software. There are shortcuts that may be taken in every stage of the development life cycle. What seems like a good idea now can result in far greater work later on. Consider these examples:

- Switching technologies may require rewriting all the software.
- Poorly understood product requirements leads to an unusable product.
- Weak architecture and design causes software to decay very rapidly.
- Poor coding practices make evolutionary design impossible.
- Lack of automated testing causes defects to hide and blocks refactoring.

Individually these issues can cause problems for a project but when combined together they can spell disaster. Review each phase of your life cycle and identify the weak spots. Figure out how you intend to mitigate these vulnerabilities.

Operational Debt The software is deployed after the development is complete. Customers begin using the software for its intended purpose. A completely different set of issues takes over. The focus shifts to how to support the needs of the live customers and keep the system healthy. The following issues become the primary focus of the team responsible for operations:

- Complex web service interactions make for difficult maintenance.
- Long release schedules force increasing project scope and lower quality.
- Lack of deployment automation causes operational errors.
- Weak system monitoring requires staff to discover operational problems.
- System can't scale to accommodate a large number of users.

After the release, operational issues dominate the mindshare. A company that may be good at developing new software may stumble during operations. Technical debt may be to blame for poor financial results and customer satisfaction issues.

Cultural Debt The last major category of issues relate to the people aspects of the business. Software is created by teams of people working together to solve problems. Some organizations excel by building a great culture while others struggle to get people to work together at all. Cultural debt can be deeply entrenched and difficult to resolve. Consider these indicators of cultural debt:

- Planning process is inflexible in the face of changing realities.
- Product plans only consider the current generation.
- Low product quality prevents any leverage.
- Lack of standard practices results in little shared knowledge.
- Lack of tools for sharing information prevents free exchange of ideas.
- No way to manage the required skill sets for the project.
- Staff is afraid and unwilling to share information with others.

These issues are all related to group norms established within your organization. These group norms are based in the cultural values that dictate the everyone's daily behavior. The culture can be changed over time, but it does take a clear plan to do so. In Part 4 we will explore in greater depth how you can create a culture that supports leverage.

Recognizing Technical Debt

Learning to recognize technical debt is a critical skill for an architect. Knowing where to apply energy and resources is a key aspect of steering the technical direction of the project. This is a place where leadership can shine and you can have a huge impact on the success of the project. Detecting problems while they are still small can prevent much greater problems in the future.

Organizational Dysfunction and Technical Debt There is a direct correlation between organizational dysfunction and technical debt. Dysfunctional organizations create products that are riddled with technical debt. Excellent teams create excellent products with high quality and high predictability. This is no accident, good teams utilize good processes that create good products. In fact, the product quality is a direct reflection of the organizational quality of the team that produced it.

Culture gets special treatment in this book because building a healthy culture is a requirement for building good software. There isn't a sick organization that can consistently produce high-quality software. If you intend to create an environment that maximizes leverage you must tackle some of the people-related issues that can block achieving that goal.

The natural order of things is to decay and experience corruption. This is true of organizations as well as the software they produce. For an organization to stay healthy it must continue to grow. New ideas and skills must be developed in order to keep pace with the business demands.

If a team fails to grow in its process skill it will inevitably slide into dysfunction. The team will lose sight of the primary project goals and become obsessed with trivial matters. This organizational debt is very similar to the decay that an unmanaged software system experiences.

Organizational debt is the result of a lack of leadership but a single champion can have a profound positive impact on a team. Perhaps you are the champion your organization needs?

When there is organizational dysfunction, each member of the team must make a choice. "Will I step forward and lead or will I duck for cover"? In the Dilbert comic strip, Scott Adams describes the Wally character who is on permanent "in plant vacation". We need fewer Wally's and more leaders who will have the courage to make necessary changes.

People can accomplish far more working together than working alone so build a team of like-minded people that can support your efforts. If something works on your team then share the successful formula with neighboring teams. Spread the best practices around and try to focus on grassroots support. Over time you can have a major impact on the effectiveness of your organization.

Example Scenario To illustrate how technical debt works in the real world, let us imagine a mythical company and how some of the project dynamics play out. The Bad News Herald is a blogging website dedicated to doomsday devotees. Bad News is founded by three prima donnas with big egos. Each privately harbors the belief that the other two are inferior, but they choose to work together anyway.

They agree on the basic idea of the business but can't seem to get to closure on the details. Instead of focusing on gaining a common vision they decide to write a complete product spec up front. One of the members thinks that this is a complete waste of time and so begins building the product that matches his ideas. Eventually he has written enough code that the other two are forced to abandon their efforts because their product ideas are incompatible.

A choice is made to focus primarily on the feature set to try and get everything up and running quickly. It is uncertain whether the core infrastructure will support the kind of traffic they may experience but these issues are pushed aside in order to add lots of advanced features to the UI. A new technology is being used that no one is really familiar with but this doesn't dampen their enthusiasm. They end up disagreeing about the persistence layer so Frank uses Redis and Bill uses Postgres. They will figure out how to exchange data between the two later.

When they begin discussing the coding tools there is more disagreement. The three developers end up using three different editing environments so sharing all of the tool configurations is out of the question. The version control process is also quite difficult because they each have a different work style. One developer adds untested code to the master branch while another maintains a separate branch that is merged every other month. This makes it nearly impossible to get a valid build of the code.

Because it is so painful to merge and build the code, the developers each begin to replicate chunks of code so that they can work independently. Eventually large chunks of the code become replicated except for the slight changes that creep in and the time between integration continues to get longer. The test cycle requires two weeks to create one build worth testing. They decide that a team of testers is required just to click on the UI. This saves the effort of writing automatic tests which requires developers. They decide the testing can be outsourced to an offshore testing firm.

The endeavor continues until finally the group has run out of resources and time - that's usually when the blaming starts. There are hundreds of decisions that are made during the course of development and every decision either raises or lowers the risk level of the project. Risk is measured in technical debt. The founders of Bad News Herald made every decision in favor of debt. They steadily accepted more risk throughout the project without understanding the eventual consequence. Projects like this often end in high turnover and high personal costs to the team members.

This unfortunate situation happens far too often. People's lives and careers are ruined by taking on unreasonable risks. There is a time to take on risk but it must be carefully considered. You must remain conscious of the level of all types of technical debt that you are taking on and ultimately, you must be prepared to pay off the debt at some point.

The Leverage Mindset   Turing Award winner Fred Brooks wrote a brilliant paper in 1986, entitled "No Silver Bullet — Essence and Accidents of Software Engineering". He pointed out that there is no sliver bullet for software productivity. No tool, language, or technology will offer everything we need in a single shot. Yet, today most engineers are still searching desperately for a panacea that will provide all of the answers.

While no silver bullets exist, there are some very effective bullets. There are best practices that produce great results every time. There are tools that are efficient and ways to manage teams that actually work. There is a body of knowledge that can be tapped into that will transform your team into a world class operation. Embracing the concept of constant incremental improvement is truly adopting a leverage mindset.

While no one thing will fix all your problems, everything you do will help. The problems are fixed by applying well-known solutions, and it is worth studying the solutions because the problems are worth fixing. Base your development practices on techniques that have been proven to work well.

A Broader Definition of Leverage Over the last three decades there has been an acknowledgment of the importance of software reuse. I recently attended a software conference session on software reuse where the speaker talked exclusively about how to package a library so it can be used by others. Reusing code without modification is the most restrictive and, therefore, the least useful form of leverage.

I think this illustrates a problem in our thinking. While reusing packaged code may save us the effort of rewriting standard functions in common use, it doesn't go far enough. If we want to use all assets to the greatest advantage we can't limit our thinking only to reusing code. The leverage mindset forces us to embrace a much broader concept of reuse and to rethink the multiple layers of understanding that can be leveraged.

We must be able to leverage:

- customer understanding
- business goals and strategies
- brand recognition
- teamwork and collaboration
- knowledge and skills of team
- project plans
- technology understanding
- architecture
- detailed design
- code
- software development tools and process
- tests and testing strategies

- release tools
- scaling information
- server management techniques and tools

By viewing leverage in the broadest possible terms, we can apply every form of understanding gained from past experience to future problems. True leverage occurs in every aspect of problem solving. We are able to recognize the common patterns and apply appropriate solutions and this yields the productivity gains that we so desperately need.

Management of Technical Debt

What if the problem before us is only similar, but not identical, to the one someone else has solved? Rarely is the problem statement identical for every one of these aspects. Everything that we do must be adaptable. Our ability to leverage is based on our ability to apply old solutions to new problems and to recognize when this produces an advantage for us. In order to maximize advantage, leverage every part of the intellectual property chain.

The leverage mindset demands that we be pragmatic. To quote Teddy Roosevelt, "Do what you can, where you are, with what you have." Flexibility is the key to leverage. We examine the current problem and consult our vast inventory of tricks to determine which one to use here. Is there a standard solution already available or should a new solution be created for this situation and used the next time? This is the leverage mindset.

Understanding is the key asset to leverage. Debt ruins your chance at leverage because it makes it too difficult to adapt. Have you every worked on a design that you knew must possess some brilliant virtue but it was rendered unusable by its flaws? The leverage mindset compels you to abhor technical debt in all of its forms because is ruins your leverage. A lot of missing tests or defects or poor structure will destroy any chance of leverage.

Leverage is the key to controlling costs so technical debt can actually be measured in dollars. Software architects need to build personal skills around having the budget conversation. "We would like to use that library but it will cost us 100 hours to replace the database driver with ours. We have 100 defects outstanding which will take 400 hours to solve and there are probably another 100 still hiding. We are at least 800 hours from release."

You have fully adopted the Leverage Mindset when you believe that your software can live indefinitely. But to make this happen you will need to align your development practices to the goal. Lots of people are looking for a "Get Rich Quick" scheme. The leverage optimization approach is to build practices that will allow us to get rich slowly. We want to maximize the long-term business payoff for every hour of work done. Aim to build high-quality software and make it last forever. Success is producing the highest value for the lowest effort and leverage is the only way that you will ever get there.

Measuring Technical Debt Technical debt is the most significant threat to every software project. Throughout the development there are thousands of technical trade-offs made. Each of these trade-offs is a technical problem that must be solved. But a decision is often made to defer solving a problem at this time. This promise to work on this issue later produces an IOU for the project.

Technical debt naturally builds over time. Each time a problem is deferred, the quality of the project has decreased. Managers naturally assume that the engineers and architects are resolving everything that they encounter. They believe that the project is proceeding with a low number of unresolved issues.

Software architects and senior engineers can recount numerous discussions with managers about the need to fix systemic problems within the design. If a manager rejects an opportunity to authorize refactoring the code, the same manager is likely to be shocked at the consequences of not resolving the looming issues early enough.

Metrics are one way that we can measure the health of our software system and the health of the development process and team. This will enhance communication and make it much easier to get approval for addressing the issues before they completely shut down the development progress.

Managers can tell when the software functionality is there. They can see the steady progress with new features being turned on. They are excited as each new build produces a pet new feature. Quality, on the other hand, is often invisible and hard to understand. The compromises made while creating the features are often unseen. It is difficult to fully appreciate the technical debt that is accumulating under the hood.

People assume the favorable state. They believe that everything is fine because that is mostly what they see. The VP of Engineering asks for a product demo and a team of engineers carefully craft a *happy path* through the product. Everyone is delighted and the VP announces the early shipment for the solution. No one wants to admit that the project is far from release due to the myriad of unresolved issues. Everyone has a natural reluctance to broadcast or hear the bad news. Thus, many projects careen toward disaster without notice.

Managing Technical Debt Quantitative measurements can replace subjective positioning. The same demo can be shown to the VP. She can get all excited about the potential of the future product, but there can also be a presentation of the current realities.

An organization that understands the metrics for success can have a compelling and accurate story about how this project compares to the performance of the current product. The story can be clear even in a highly political environment because problems can be revealed in the actual data.

People will optimize the things that have the most value. People have an innate desire to win so clarifying what success looks like is a great motivator. The metrics that we collect should either directly reflect success or lead to a calculation that does. For example, revenue is a direct indicator while number of modules must be part of a larger picture.

Reward the things that you value most - people will automatically learn and adapt to the desired behavior. If you have a highly visible dashboard many people will experience a competitive desire to play to the dashboard.

This leads to the metrics dilemma. Be careful what you wish for, you just might get it. If you measure the wrong thing you will get it even if it is actually harmful to your success. I recall a Dilbert cartoon where Wally learned of a $100 incentive to engineers for every defect they discovered. Wally says, "I'm gonna write me a minivan!"

Best Practice #2 - *Measure and manage technical debt weekly.*

Problem Technical debt occurs when we take shortcuts. There are tasks that must be completed on the project but we have neglected or delayed their completion. The decision to delay tasks may, or may not, have been conscious but either way it is an IOU against the project. Some systems are so riddled with debt that the net worth is negative - it would require more to fix the system than it is worth to the business.

Most projects have no way to measure their technical debt. Debt naturally accumulates and requires a concerted effort to eliminate. A poorly constructed system can decay and fall into disuse within 18 months of its release due to crippling technical debt.

A large technical debt will block all of the possible leverage for follow-on projects. The team itself may experience very high turnover due to loss of confidence in the leadership, resulting in all of the project knowledge being lost. A project with no leverage can easily cost ten times the expected budget. A product with excessive debt may even severely damage the company.

Solution Eliminating debt requires being able to measure it. There are many different types of debt. A dashboard can track the most important debt terms in your equation. Debt is measured in hours, which can also be translated into monetary terms, if desired. The total debt represents the unfinished business on your project and is the sum of all the terms.

The software development process itself is an exercise in debt reduction. At the end of the project you should have less than 10% debt. A project that ships with 50% debt is really only half done and will fail after a brief time in the market.

Next Steps

- Complete a self-assessment of the types of debt in your current project.
- Select the top three areas of debt that you intend to decrease.

- Find two other people that share your passion.
- Create a simple and actionable objective for each goal.
- Assess your balance (Function/Quality, Visible/Hidden).

Chapter 3 - Best Practices = Leverage

> "A best practice is a method or technique that has consistently shown results superior to those achieved with other means, and that is used as a benchmark."
>
> ~ *Wikipedia*

Developing Best Practices

Best practices must be developed because they aren't a natural byproduct of creating software. It requires a concerted effort to build and share software processes throughout the organization. A significant part of an architect's mindshare should be applied to how development should be done, not just producing the next product.

Leverage is the best way to control the cost of developing software. However, the flaws within our organizations and our software may block leverage. Technical debt prevents us from reusing the knowledge that we already have to produce new products. So how do we escape this dilemma? By developing a standard set of practices over time. This leaves us free to fully leverage the solutions that are already proven.

Our primary objective is to produce the maximum amount of software value for the lowest cost. If software is simply a repeated act of problem solving, then the cost of the software is related to the number and complexity of the problems that we choose to solve.

Pay Off Technical Debt With a heavy load of technical debt, each problem that we attempt to solve is difficult. The remaining defects, missing tests, inappropriate data structures and interfaces, all work against us. It can become quite difficult to even read and understand the code. In the extreme cases, it can become virtually impossible to maintain the code. Have you ever been on a project where there was no one available with any knowledge of how to modify the code successfully? At that point the code is no longer maintainable. The cost of maintenance is now higher than the cost of replacement.

There are techniques that will prevent this situation from occurring and the same methods can be used for restoring health to bad software. They can be learned and applied by any developer.

Technical debt is the problem and implementing best practices is the solution. Each ailment has a remedy; each problem has a solution. A best practice is a recipe for solving a certain problem by applying a known solution.

Capture Recurring Solutions As we explore the problems and solutions we'll offer tips for building your own catalog of best practices. Consider the catalog offered in this book as a starter kit on which you can build upon. Use the ones that work for you directly, modify others, and reject the ones that don't make sense for you. Going through this tutorial will sharpen your awareness of your team's current software development practices.

As you build your own catalog, share your ideas with your team and with other people too. By learning and teaching best practices you can multiply the positive impact within your company. Most organizations are in desperate need of this type of leadership.

Catalog of Best Practice A central repository for standard solutions and best practices is essential. Eventually you can build out a large system to track all of your practices but this will take years. It is best to start small with a simple solution.

Each practice can be documented with a simple text template for the key aspects of the practice:

- Problem - What is the issue that is being solved? When should this be used?
- Solution - What are the key elements that need to be applied?
- Discussion - What are common issues, variations, and limitations?

Start with a very simple template to encourage engineers to write the practices down. Remove every excuse and barrier that might prevent someone from contributing. Once a solution is identified for a particular problem it should be the standard answer for all future problems of that type. It is important for the best practices to be shared and universally applied. Otherwise, you are putting effort into a system that generates no value to the business.

If someone uses a different solution you have three possible responses:

- Add the new solution to the catalog
- Modify the existing solution to include the new twist
- Replace the approved solution with the new one

Best Practice Collaboration   The system must be easily searchable. A text search and a tagging system makes it easy for anyone to find what they are looking for in less than a minute. Tags are better than folders since there may be many ways to retrieve the same information.

The system should meet the specific needs of your engineers. It might be integrated in with your source code or built as a stand-alone application. It should live independently of any specific product.

This system should house all of the latest process information that will teach your team how to do everything. Think of it as a cookbook of tricks that can be applied whenever there is a need.

A system to track best practices is a great way to improve your overall software development process. Pockets of excellence can be spread throughout the group so that everyone can steadily increase in skill. Software process knowledge is a key asset in your organization. Measure the size of your system - it should be growing over time. Measure the growth of different aspects of your system:

- articles
- words
- authors
- technologies
- code snippets
- tools

This central repository of best practices represents your health as a software organization. Engineers should be rewarded for contributing to the shared base of knowledge and they should also be reprimanded for refusing to share their knowledge. Knowledge hoarding is very destructive.

Promoting Software Process

An organization based on continual learning will thrive over the long term. Jim Rohn said, "Success is nothing more than a few simple disciplines, practiced every day". Best practices take a long time to develop, but once learned they should be shared as widely as possible.

Understanding how to build software effectively is one of the most important assets you have. Build this body of knowledge with each product generation. Make sure that each team member is aware of the work to build standard practices. When every individual is fully engaged in building your best practice inventory rapid progress will be made to grow expertise across the entire organization.

Learned and Shared Start with a simple one page description of how to solve a problem then let multiple people enhance it until it is the definitive word on the topic. Build the documentation as you build the knowledge. Complex tasks will require more description. In this way, you are amortizing the cost of learning new skills and building an asset that may last for many years.

It is wonderful to be looking at a difficult problem and discover that someone else has already solved it. Think of the benefit of being able to leverage a working answer within an hour when the expected effort was several hours of work. Be generous with your coworkers - leave them a nice solution that will save them a lot of work. You want to participate on both sides of this equation.

Not all solutions are equally good. It is important that everyone is allowed to edit the repository of software processes. This provides an environment to mentor the less skilled developers. You may want to consider adding a voting mechanism into the system, similar to the one that Stack Overflow uses. This gives certain ideas greater weight.

Your most skilled engineers should constantly be curating the content of the system because you are telling all engineers to apply these solutions.

The senior engineers can verify that the solutions are appropriate. Over time, you will build a set of ground rules about who can edit what content. However, at the inception be light-handed - there is a bigger danger of the system falling into disuse than there is of anarchy.

Make contributing practices to the catalog fun. Announce when new contributors have created their first posting. Make contributing a worthwhile activity.

When developing new solutions, the first idea is not always the best. Some ideas require multiple attempts to finally create an elegant solution. The process recipes in the catalog should withstand scrutiny and review. Experiment with the ideas to either improve or replace them. Periodically review the catalog and assign engineers to investigate the processes that cause the most problems.

Experimentation is the key to developing great solutions. Reviewing a problem can reveal fresh insight and lead to a better solution. Don't be hesitant to replace any solution with a superior one - but be sure to test it adequately over the entire range of possible environments.

Standardizing Practices Standard solutions to common problems will lead to predictability. This is one of the primary goals for standardization: anyone can apply a solution with a high confidence of getting an acceptable result.

By replacing manual steps in the solution with automation you can increase the repeatability even further. The computer will execute the steps in exactly the same way every time. This removes the potential for human error. Of course, the automation must be thoroughly tested. One advantage to automation is that it is easy to build the verification directly into the script. For example, if a process is intended to clear out a directory you can check that there are no files present when the script ends.

Remember, no panacea exists for software productivity and quality. And yet, we keep looking for that one magic tool that will make us fly. Perfectly reasonable people are turned into fanatics when it comes to tools, languages, frameworks, operating systems, and text editors. But the reality remains - no one tool or process will meet all our needs.

Although some tools and processes are clearly superior to others, the gain that we receive is often only a small part of what is promised. A best practice is any solution that gives us an advantage over the way it used to be done. Stock your catalog with solutions that work. They might not be silver bullets but they can still slay the dragon. Use the best ammo you can afford and constantly develop better ammunition.

Repeatable Results Best practices should be developed incrementally. Every process already is being done in some way today. Start with a short description of the current process. What steps are being done and what problems aren't adequately addressed? That is the baseline - build from it and only change one thing at a time.

Avoid the temptation to scrap everything in favor of a new fad. You are already building software with some degree of success. Use what works and look for improvements in the areas of greatest pain.

Move slow and carefully in order to move faster. Change one thing and study the results. If you find that you have made a mistake then undo the new process. Once you have identified a clear improvement - use it everywhere. Go through your entire system and replace the old with the new - don't let multiple solutions live together over the long term! A temporary, transitional state is acceptable but multiple solutions to the same problem create confusion with new developers and should be unified.

A winning solution should also be applied at every future opportunity. There should be a clear difference in circumstances that should allow an engineer to choose between two similar solutions to a problem. Part of the documentation for the solution should include a description of the context in which one solution is better than another. Try to avoid situations where there are competing answers that amount to a matter of personal preference.

Championing Best Practices Best practices are a vital asset for your organization and can be as important as the product itself. Software process is the golden goose that will allow you to create great products in the future. This may seem obvious but many groups fails to grasp the tremendous importance of good software process. Engineers commonly find themselves under pressure to take shortcuts in order to meet unreasonable schedule demands.

The sad truth is that failing to invest in process will only worsen the schedule crisis. Taking a shortcut will hurt you every time. Not right away, perhaps, but it will bite you eventually and it usually happens when you can least afford it.

Part of the value that software architects provide to a project is the judgment of what is good for the long-term health of the software. Evaluate the true merits of other solutions and be willing to acknowledge the advantages. Only then, are you in a good position to champion an alternative. Your confidence should be tempered with humility - people that insist they are always right have little credibility with others.

Spheres of Influence: Me, Us, Them

Ideas and solutions can be found in a lot of different places. Best practices must first be leveraged at a personal level before they can be leveraged effectively at the team level. At the broadest level you have opportunities to learn from the processes of other teams, groups, and companies.

The first circle of influence is personal, how I develop software. What best practices govern my development activities? What problems do I currently understand well? Do I leverage these best practices to the fullest extent I can? As you gain traction in each sphere of influence it is time to leverage what you have learned to the next larger group.

The role of an architect includes a mandate to exert a positive influence on the larger organization. Your impact must be felt far beyond your immediate work group in order to be successful at your job. The leverage mindset can have a profound impact on an entire company. The goal is to prove out ideas and then to influence others to adopt them for the maximum benefit.

Personal Practices Before I can contribute anything to my team I must perfect my own craft. Here's an opportunity for a quick self-assessment - begin to consciously observe how you develop software and the techniques you use. Awareness is the first step toward improvement.

Figure 4: Spheres of Influence for Best Practices

Now make a list of standard processes that you frequently use. Think about how you might teach these skills to a new member of the team. Teaching is the best way to learn deeply so this step might p be very valuable to you.

Consider investing a couple of hours each day to improve your craft. This, after all, is real project work. I guarantee that once you begin this practice, it will become the most critical part of your work life.

Don't wait for permission to improve your skill. This is your investment into your career and will benefit both your current and any future employer. This type of effort is required to stay current in our fast-moving industry. You should be able to look at your learning agenda and see how you have invested the 400 hours over the last year. A one week annual training class is completely inadequate for you to build all the skills you need.

The best way to document best practices is as a catalog - you are providing recipes for a cookbook on how to build software. The book, "Design Patterns", does an excellent job of demonstrating techniques for documenting many types of patterns that can be used by others.

Keep the requirement light in order to make it easy to document new practices. You can always make it heavier once the system is in mainstream use. A best practice should address three aspects:

- Problem - recognize conditions where this pattern might be beneficial.
- Solution - recipe for creating a successful implementation.
- Discussion - adjustable customizations, limitations, and application notes.

Work to master your own engineering practices before trying to export them. If you try to share processes that aren't fully developed you could damage your credibility.

Team Practices The next level of process development is to leverage the personal learning to benefit the rest of your team. You have built a repertoire of tricks and now you need to share them with your team.

You can also benefit from the tricks others are using. Success attracts its own publicity - find things that work and tell others why they work for you. They may be willing to apply the idea and give you some feedback on how to improve it.

By making it easy to capture process information you can build your own catalog of best practices, custom-tailored to the specific needs of your team. Efforts to capture new ideas can be ruined by placing too many demands on authors who intend to submit new best practices. Your initial goal should be to capture the actual practices of your team. Once you have a good inventory you can go back to the authors and seek to increase the quality. By then they already have an investment in the ideas and will be willing to make the needed improvements.

By using this process the idea will be authored by several people and take on several new champions as a result. Share your system for organizing the best practices and invite collaboration. Consider hosting a lunch time seminar or webinar with your coworkers in order to show them the recent progress. Help people understand the benefits of contributing to the Best Practices system.

A concerted effort of process improvement will take time. Encourage people to experiment with the ideas and make improvements. A great practice is produced after many hours of experimentation and thought have been applied.

Where there is pain, there is an opportunity for improvement. If everything is running fine there is no reason to dedicate effort to make it better. The more stress that is in your project, the more improvement can be justified. Let the reality of your situation dictate the level of energy to be invested in process improvement.

Michelangelo once commented that creating a statue from stone was simply a matter of removing the extra rock to free the statue that was already inside. I like to think about software that way. An app wants to come forth and it will unless something is blocking it. A best practice is nothing more than removing a challenge that would mar the artwork that is our software. Your next product might have already been released were it not for the thousand problems that prevent you from releasing it. Best practices accelerate the transition from lump to art.

Industry Practices Now we are ready to set our vision beyond our own team. We look at how other groups have solved the same problems that we face. It is important for us to start by trying to solve problems on our own - only then will we fully appreciate the attributes of the solution. Software development occurs in a variety of situations so it is impossible for a general purpose repository of best practices to be universally applicable.

No one else has your exact problems so your storehouse of best practices should meet your unique needs. Make full use of the similarities of processes that you adopt from others but also respect the differences. You can safely ignore everything that does not apply to your situation. Be sure to adapt the ideas rather than just adopt them. You will learn a lot as you try to figure out what actually applies and the end result will match the specifics of your business needs.

You may incorporate practices from others but the authorship and review is your responsibility. Create articles that document each practice and try to limit the text to a single page initially. Make it easy for people to include information that they find on the Internet but ensure that each idea is adequately reviewed before it is put forth as a best practice.

Every team has ongoing challenges. Create a list of the significant problems that you wish to have solved. Prioritize this list and make this "10 Most Wanted" list the focus of your exploration. Assign your most talented engineers with a charter to fix these problems. These are the problems that are keeping you awake at night - finding great solutions will make everyone sleep better.

You can even have some fun with it by running a competition for the solution to the nastiest problem. Select a real prize to sweeten the pot and align the team around the benefit of solving your most important problems. This goes a long way toward building a shared sense of identity. "We are the team that pulls together and gets things done."

Seek to standardize on the processes that have already been identified to be best practices. Engineering practice is more than a matter of preference - developers should use the best possible process for any given task. Learning a new way to do something that already has a standard solution is a waste of resources. However, there must also be room to refine existing practices and add new ones.

A standard practice should be applied universally in a specific context. If our team chose Git for version control, someone would need a really good reason to use Subversion. The burden of proof lies on the engineer that wants to promote a different practice - they have to demonstrate a significant benefit to warrant the energy necessary for everyone to change. Allowing each engineer to select their own tools and processes at will produces a lot of waste within the project. The team must be open to review and improve the practices but once approved, they shouldn't be ignored.

Essential Practices for Leverage

Throughout this book we offer best practices that you can apply immediately to your project. These recommendations will be presented in the chapter that covers the related issues in depth. Here is your starter kit of best practices.

Essential Best Practices The Leverage Principle

- 1 - Make leverage your primary goal.
- 2 - Measure and manage technical debt weekly.
- 3 - Capture, develop, and standardize best practices.

Development

- 4 - Select technologies that will support your leverage goals.
- 5 - Create components with strong encapsulation and standard interfaces.
- 6 - Use a balanced approach to development resulting in minimum code complexity.
- 7 - Use diff testing to generate maximum test coverage.

Operations

- 8 - Build for continuous delivery of software and use end-to-end automation.

- 9 - Build your system from loosely connected services for optimal flexibility and reuse.
- 10 - Automate everything except the decision to deploy.
- 11 - Monitor everything that you care about.

Culture

- 12 - Create a robust system for sharing all types of knowledge.
- 13 - Build healthy teams by investing resources, creating a team manifesto, and tracking team goals quarterly.
- 14 - Make learning a top priority by measuring it, planning for it, investing in it, and rewarding it when it happens.
- 15 - Adjust plans throughout development to capitalize on new learning and track progress weekly.

Best Practice #3 - *Capture, develop, and standardize best practices.*

Problem It takes a lot of time to properly solve a business or technical problem. Once an optimal solution is invented it should be used to the fullest extent possible. This can't happen unless there is a standard method to communicate the problems and solutions.

If there isn't a system to track best practices then each time an engineer runs into a common problem they must solve it on their own. A lot of wasted learning goes into solving problems that have already been solved by someone else. This is really an organizational failure to preserve and utilize knowledge. Standard solutions should be readily available for common problems. This prevents several people from spending time solving the same problem, over, and over again.

In addition to the direct time savings, there is also a quality issue. A much better solution can be developed if there is a high likelihood of using it many times. A single developer, intending to use the solution once, may choose to add artificial limitations and constraints. It is better to have a well-defined solution and apply it repeatedly in every appropriate situation. This allows the solution to be designed, reviewed, tested, and tweaked.

Training can help individual engineers avoid the common pitfalls of misapplying the solutions. Without standardizing on best practices there is a significant loss of leverage and unnecessarily inflated costs for the project. The unintended result is to force everyone to be an expert and yet, fail to reward them for it.

Solution A simple system can be created to capture all of the best practices in one place. Each problem and solution is clearly discussed so that it can be applied to future situations. The initial description should be simple and only discuss the core concepts of the practice. This makes it simple to document new practices. Simplicity is the key to promoting adoption of the system. The first discussion should be limited to about six paragraphs and explore only the essential aspects.

Later improvements will add details to the initial description - after there has been a chance to validate the basic ideas. The catalog will grow in both depth and breadth as new ideas are added and developed further. Tags will provide easy access and organization to the best practices and will allow ideas to be retrieved from multiple groupings. A text search will also offer an easy way to retrieve the content.

Next Steps

- Review your current practices for gathering and sharing best practices.
- Make a list of the best practices that you already have documented.
- Create a list of new best practices that you should be tracking.
- Select the three most important practices and a simple plan for improvements.
- Identify and recruit potential allies for process improvement.

PART 2 - Leverage in Development

Figure 5:

Chapter 4 - Technology Leverage

"If you do not change direction, you may end up where you are heading."

~ *Gautama Buddha*

Technology Provides Leverage

Software development is about solving problems. Sometimes this requires building an entire solution from scratch, but more often your solution will be built on top of existing technology that solves many of the problems for you.

Technology selection is really a form of leverage - you are using existing software to produce a solution without having to build your own software for the same purpose. A single tool may reflect the knowledge of thousands of hours of careful thought and development. When you build your own tool without considering an existing solution you may be duplicating problems that others have solved well.

This chapter will focus solely on how to optimize the selection process so that every project gets a boost from the use of a great tool set.

How Technology Selection Affects Leverage Every solution that we build is based on some supporting technologies. This makes decisions about technology selection critical to the success of every project. Many organizations fail to give the technology selection the attention it deserves and some decisions are made arbitrarily.

Choosing good technology can give a project an extra boost but a stable set of technologies is also important for ongoing productivity. Switching technologies too often can have a detrimental effect on the project because of the learning curve involved. The reverse is true as well - I have seen the negative effects of continued use of old technology when new tools could have provided a significant boost in productivity. Leverage requires careful management and timing of technology changes.

Standardize on Great Tools When you find a suitable standard solution - use it as much as possible. Only build your own technology when you can't find a workable solution from somewhere else. I estimate that building your own tools will cost you roughly ten times the amount of using existing tools.

Fill the gaps in your tool set by being intentional about looking for new tools. Make a list of the problems that you have yet to solve and then look for tools that have already solved this problem well. As you find a great tool for a problem on your list, add the tool and communicate that selection with your team. An architect often has to think more like a consultant and less like a programmer when approaching tool selection.

Tools enable leverage, either directly or indirectly. When you write your own software, you must create your own leverage. First you define the simple solution and then you expand it to a more general case. A tool might be able to solve your simple problem and it is much more likely to get you to the sophisticated solution even more quickly. For years, every team I worked on wanted to build an issues tracking database. This offered no value to the project and consumed a lot of effort.

Your company should have a system to manage the common knowledge that engineers can draw on when solving any problem. Technology selection, usage, limitations, cookbooks, processes, and tutorials should all be available to everyone in your organization. An engineer should be able to quickly query and obtain any information related to any problem. This system should offer the definitive answer to each technical problem.

Applying a well-known solution is easy but as you begin to use new tools you must understand the assumptions that are embedded in the tool. Junior engineers often misapply powerful tools with disastrous consequences. There is no substitute for truly understanding what the tool is doing. To have a beneficial result you must know the limitations and circumstances required for success. A chainsaw is a great tool in the hands of a skilled craftsman but can create a disaster when wielded by a novice.

Getting the Most from Technology When selecting technology, be intentional about your choices and trade-offs - don't make the decision lightly. Create a team of reviewers and work toward a consensus on the tool selection - don't try to rush the process to closure. The following technology attributes will mean the difference between success and failure.

- Flexibility - How adaptable is this process and when does it break?
- Scalability - Does the solution work at different sizes?
- Extensibility - How can the process be adapted to new situations?
- Dependencies - What are the key linkages to other processes?
- Learning Curve - How difficult is it to get people properly trained?
- Usability - Is it difficult to use even after thorough training?
- Reliability - Can you count on predictable results or is it hit and miss?
- Localization - How can the process be adapted to more specialized usage?
- Performance - How can the efficiency be improved to yield a higher ROI?
- Longevity - Is this a sustainable process?
- Culture fit - is remote access, 24/7, and degree-of-co-creation supported with tool?
- Leveragabilty of the tool itself - is it customizable? Does it solve 80% of the problem while truly only needing 20% effort to have the tool fit the need?
- ROI of learning and cost of using the tool - is it lowered by repeated usage?

Recall our earlier discussion of NIH Syndrome (Not Invented Here). This functional disorder results in engineers compulsively reinventing solutions that already exist. There is usually a case made for some attribute that is missing from the existing solution. If we build a solution it must always be in line with the true economics of the situation. This decision should always be justified by creating a higher ROI. Save your invention for the real problems that must be solved.

Technology Trends

Technology is constantly evolving and changing and many of these changes can have a dramatic impact on your current product plans. Study the trends in technology to keep current with changes as they occur. A software architect should spend about 30% of their time learning and experimenting with new technology. This investment is required to intelligently inform the constant decision-making that occurs throughout the project.

There are two trends that are expected to have a substantial influence on how we develop software. The trend toward open source software has changed the value proposition and cost equation on most projects. The second major technology trend to watch is the move away from isolated software services toward complex integrated web services deployed in the cloud.

Open Source Software At the turn of the 21st century technology selection was about choosing between buying commercial products from established vendors or building your own tools from scratch. The tools on the market were of high quality but were very expensive. Now there is a viable third option. Over the last decade we have seen the rise of amazing open source software platforms. These tools are available for use without any cost. The very presence of open source software has driven down the cost of commercial software. This is very good news for developers! The cost of fully outfitting an engineer has gone from thousands of dollars to a couple hundred dollars.

The quality of open source software has risen to the level of commercial software. It is easy to assemble a team of people that use only open source tools and training is readily available to bring new teams up to speed. Even though open source software is now a safe path forward, many established companies are reluctant to embrace open source tools because of the potential legal entanglements. The Patent Trolls have left an unnecessary legacy of fear regarding open source software. By understanding the licensing agreements before working with any new software you can include open source tools and commercial software in your selection process. Most agreements now are very open and don't impose any unreasonable restrictions on the use of open source software.

Contributing to open source projects may also be a way to get the specific functionality you want without having to reinvent the entire system. You can build your own customizations on top of a standard platform and keep them proprietary. Or you can choose to contribute your customizations back to the platform. Many large companies have embraced open source software as a strategic extension of their own development capabilities and they are assigning employees to make specific contributions to open source projects. This allows the company to get specific custom benefits and allows a contribution to the wider developer community.

Working with open source projects is simple. Just fork a repository in order to make custom edits to open source software. Then if you want to contribute your changes back to the base repository you generate a pull request. The maintainer of the base repo can then evaluate your change and incorporate it back in if they like it. There is no way to force an update into an open source project. Most projects have stringent requirements that prevent low quality contributions. If you wish to contribute to a project make sure that you are prepared to put in the effort that is required to clear the bar.

The decisions to use open source software and contribute back to the project are two independent choices. Even if you don't have any interest in contributing to a base repository, consider using open software. Watch the trends - the future of technology lies in the open source direction.

Cloud Services The second major technology trend is that most software is now executing remotely - made accessible as services. Web applications and web services provide all kinds of value. The days of the server in a closet are over - the era of DevOps has arrived.

Existing solutions can typically be accessed by getting authentication credentials to a hosted server that is running in a remote data center. Even when you are hosting your applications within your own company data center these applications are very similar to the ones running at Google or Amazon.

One concern when buying software solutions is to beware of proprietary tools and the dangers of vendor lock-in and excessive licensing fees. A single vendor can hold you hostage with no recourse if you don't have any reasonable alternatives.

The world is constantly changing and the software must also be constantly changing to keep pace. This is true of tools as well. Some people expect to buy a tool and use it forever, but this is seldom reality. Software has a freshness date. When you choose a tool think about its future as much as you evaluate its current state. Where will it be in two years?

Software is in a natural state of steady decay. This is true of your product and also the technology stack that your product is built on. Don't choose any technology unless you believe that it will be around in ten years. Stay away from experimental tech until it is fully proven. Keep to the mainstream rather than finding innovative eddies.

If you lose confidence in a technology platform then consider how you can move off of it. Fixing problems in a failing technology is a losing proposition. Never bet against a trend - all new tech selections should go with the trends.

Embrace Standards Many groups resist using standard tools. They feel that inventing custom solutions is required even when it produces no benefit to the product or business. Use standard development processes, tools, protocols, and solutions when you can. The only justifiable reason to implement custom software is when no suitable solution already exists.

Continuously investigate new technologies but do it in parallel with product development. When you are first learning a new technology it may seem almost perfect. The marketing promise is that it will solve all known problems and produce effortless weight loss and straighten your teeth. It is only when you have worked with a technology for awhile that you can actually understand its limitations.

When you do decide to deploy a technology for the first time, do it as a pilot project. Pick some low risk area of your project and try out the technology. Don't try new technology in a part of the project that is already carrying significant risk.

Risk means that there is some probability of bad things happening. If you can't absorb the risk, don't try out new technology on a large scale. After you have proven that the technology can be used successfully the risk is far lower. Now you can use the tools in a broader way throughout your development. Even then, be wary of scaling too fast - one misfire can cost a lot of credibility and time. If the technology supports this, sometimes it may make sense to use parts of a tool/process in a phased approach as a way to reduce risk.

Technology Investment

Get the most out of your technology investment by making careful decisions based on your product plans. Assemble a small team of people to guide your technology in the appropriate direction. These decisions are too important to leave to a single individual. Match the business plans to the existing technology to look for gaps.

Balance the desire to use cutting edge technology with the understanding of the true cost of transition. Evaluate the age of the technology that you are currently using against the competitive landscape. Gather all of this data together to make an informed choice about how to serve the business goals. Remember that the primary goal is to maximize the leverage of your efforts.

Technology Costs There are three real costs of using technology:

- Cost of software licensing
- Cost of training
- Cost of operations

The licensing fees are the most obvious cost so this is typically scrutinized thoroughly because someone will actually write a check. However, the other costs will dwarf the licensing cost for most systems.

The training costs are associated with getting either end-users or developers trained to make effective use of the system. For systems that affect a large number of end-users, it is important to do a business evaluation to make sure that the software will meet its intended purpose.

If the system is used by software developers as a part of a larger system then the issues are very different. Software technologies must be integrated together with custom technologies so that the new system can be created.

New software development tools carry a large learning curve. Learning a new technology comes at a high price. It will take the typical engineer about a thousand hours of practice to be truly proficient. Multiply this cost by the number of engineers and the fully loaded cost of one engineer.

This can easily account for 50-70% of the entire project budget. The productivity of the fully trained team must be phenomenally high to make this worthwhile.

The third cost is associated with the ongoing operations of the software after all of the end-users and developers are fully trained. It is easy to calculate the cost of the existing system. But be wary of the marketing claims that the new system will solve all of your existing problems. Switching to a new technology typically brings a solid benefit in the long term, but the benefits are often overstated and the learning curve is often underestimated.

Technology Portfolio   There are two primary dangers in technology selection. Some companies only use technologies that are mature because they are viewed as safe. It is easy to find many developers who are fully qualified to use mature tech with high productivity. These companies may grow comfortable with familiar technology. They risk being out of touch with what is happening in the broader industry and as a result release products that are increasingly irrelevant in the market.

In contrast, there is a danger of making repeated investments in technology that isn't yet proven and may be unstable. This can make it difficult to get the product to release. An inordinate amount of time is spent either learning the new technology or fixing underlying bugs in the infrastructure. You should never ship a product using tools that are still in Beta.

There are three maturity levels for technology based on how established they are in the industry. Each level has its own unique advantages and limitations.

Established Technologies - These technologies have been mainstream for at least ten years. You can find tens-of-thousands of developers that are well trained in using this tech. There are a hundred technical books that cover everything that you can imagine about how to apply the technology. The limitations of these technologies are well understood - the tool isn't glamorous but it gets the job done. Many engineers will ride their entire career out working on one of these mature platforms.

Emerging Technologies - Tools are always being created to yield breakthroughs in productivity and quality. Of all the tools created every

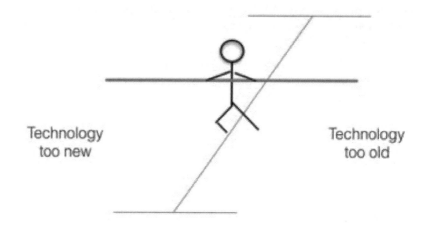

Figure 6: Balance the use of Old and New

year, there are some that have such a benefit that they begin to gather mainstream attention in the industry. This is the solid area for new ideas that are rapidly replacing the old ones. A productivity boost of two times isn't uncommon using emerging tech compared with the old (Established) technology.

Experimental Technology - This technology is so fresh the paint is not yet dry. Engineers on the cutting edge find experimental tech exciting but it is quite unproven. This type of technology often promises productivity ROI ten times greater than the established technology. You can play with this tech in a confined lab space but don't make it the basis of your business. That would be a high risk move - if you bet your next product on the use of this technology you may lose everything.

Balance the Portfolio Technology goes through an adoption cycle. It progresses through the different market stages: Innovator, Early Adopter, Early Majority, Late Majority, and Laggards. Each technology also goes through stages of maturity as it develops. Your company's portfolio of tools should directly reflect your company's business goals and risk profile. Let's consider these stages in regards to building a tech portfolio.

A Tech Startup will have a very different profile of tools than a Fortune 100 giant. Each team has a different sensitivity to risk and views risk in different terms. Develop your own profile to help make decisions about how you invest your time investigating tools at different maturity levels. Here are a couple of portfolios that may work for you but you'll have to calibrate these for your organizational preferences.

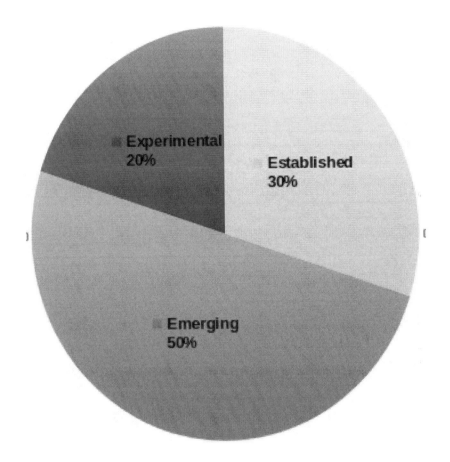

Figure 7: Startup Company Portfolio

Technology decisions should always be made beyond the needs of a single

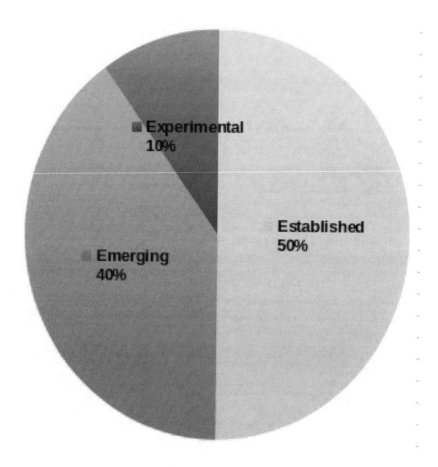

Figure 8: Mature Company Portfolio

product. Create a road map of business solutions and technology needs you foresee in the next several years. This road map should be constantly reviewed and tweaked rather than created and forgotten.

Often large companies are too reluctant to embrace new technologies until there is a crisis. Then they dive head first into new technology that is unproven. This is a recipe for disaster. If you find yourself in this situation, allow yourself time to evaluate the technology in a pilot project that doesn't put the flagship at risk.

Road Map for Technology, Tools, Process

Building a multi-year technology map is a critical planning tool. What key technologies will you be investing in over the next three years? How do you plan to phase in platform changes? Build your technology plans in parallel with your project planning.

Make your technology choices so that they give you the highest possible leverage opportunity. Build a business platform and a technical platform from which you can launch multiple generations of products. Plan for how your business will scale over time. How will you change your operations to meet future needs? Make sure that the technology that you need is already in place by the time you need it. Create a five year vision and make sure that you can achieve the scaling you need.

You must guarantee that any new technology choices are an improvement on what you already have. This starts with understanding all of the characteristics of how your current system behaves. Clearly identify the requirements that need to be met by switching technology. Try to estimate the amount of technical debt that these deficiencies are costing you in hours. If you have a numeric cost for the debt you can estimate the benefit after you pay the cost of conversion.

Be sure to weigh both the cost and benefit at the same time with the same scale. Otherwise you won't end up making a good business decision. Without accurate estimation you will end up relying solely on the personal preference of the person with the most power - this is rarely optimal.

Build a Road Map There are legitimate reasons to switch technologies. There are times when a current technology no longer meets the business needs. When this happens it is time for a change. A road map brings stability to your technology choices by balancing the short-term and long-term needs of the business.

Don't switch out a lagging technology (Established) with one that is not quite ready for use (Experimental). Only select technologies for your mainstream development that have a proven track record. Everything looks good in the marketing pitch.

Insist on a hard-nosed business case - the new technology should always be faster, cheaper, better, and more scalable that what you are currently using. Don't be swayed by the marketing hype - prove that the technology is a good fit for your business goals.

It is better to use a technology for too long than to use it for too short a period of time. Replacing technology has a lot of hidden costs and until the current situation is costing you a lot of pain, stay with what you have. You already understand the limitations of your current technology. You know how to put up with all of the quirks and idiosyncrasies and your engineers understand how to work around many things.

Every technology has a learning curve. A minor technology change will cost each engineer about 100 hours of learning. A major technology change will cost about ten times that.

I was on a project once that switched from Microsoft C# to TrollTech Qt. We had a huge code base that had taken years to build. We had about 300 engineers on the project that were highly skilled at C#. The learning curve cost the company 300,000 hours of productivity for very little business gain.

Any technology change must be justified by the cost savings it produces. Don't fail to calculate the true cost of educating your staff or you will make some very costly mistakes.

Manage Technology Transitions Make a long-term commitment to your plan. If you can't commit to a technology for several years then postpone the decision until later. In this area, you are far better not making a decision that making a hasty one.

A major technology switch will never pay off on the first project - it will likely take two years to break even. After three years you will begin reaping huge benefits from backing the right technology.

Switching technologies too often will undermine everything else you are trying to do. Wait for a clear choice and then support it and let it play out. Once you have committed to a technology, use it for everything. It shouldn't be up to individual engineers to decide what they want to use - this is an organizational decision.

Larger organizations have many projects going on at the same time. You will need to stage technology choices so that they don't adversely affect the progress on current projects. Any new technology should be tried first on a small isolated pilot project before it is rolled out across the entire product line. This gives people in your organization a chance to thoroughly vet the technology and gain confidence in it. This experience can then be leveraged as the tools are rolled out on additional projects. This staging gives you an opportunity to manage the learning curve required by the technology. The engineers on the pilot project can act as trainers and mentors for the others.

Essential Tools and Process Many engineers believe that tools are a matter or personal preference. It is important for your organization to have a clear policy around how tools are being used. There is a strong need to create a set of best practices that your team uses to collaborate with one another. Otherwise it becomes very difficult for engineers to work with each other.

Strike a balance between identifying great tools and standardizing their use and giving the engineers some latitude in how these tools are used.

Here's a list of categories of tools to include on your roadmap:

- Web application framework
- Web test framework
- Unit test frameworks
- Code editing with refactoring tools
- Code review (like Gerrit)
- Static analysis

- Version control
- Issue tracking
- Code metrics
- System configuration and deployment
- Virtualization software
- Hosting services
- Communications
- Documentation and knowledge management
- Packaging for reuse (NuGet, PCLs...)

Give special consideration to the following questions:

- How do you select technologies?
- How often do you change?
- What are the reasons that would cause you to switch?

Core Processes Build standards for all of your core processes. Think about how you could rebuild your entire business from scratch from a basic set of artifacts. Imagine that you have to document everything that your organization does so that it can be duplicated by a completely different team.

A core process is any knowledge that is required (essential) in order to do a job to a minimum standard. Under no circumstances should this critical knowledge exist only in the brain of one person. The mental exercise of a "Hit by the bus" scenario is useful here. What would you do if this person didn't show up for work one day and this core knowledge was gone? It must be captured and distributed as shared knowledge that belongs to the organization and can be accessed and replicated as needed.

I have worked on projects where it was virtually impossible to build the software from source code two months after it was released. In this situation, the organization was under a legal obligation to support the software for fifteen years. No one in the organization felt that it was justified to do the extra work to make sure that the source code could be rebuilt. Most readers can certainly identify similar failures on your projects. These large scale lapses don't happen overnight - a long series of minor shortcuts ultimately culminates in disaster.

Any software development company should function with integrity in regards to its own products. If you have an essential process that isn't managed properly it can act like a hole in a boat. If you only have one small hole you may be able to manage it. But most people would agree that boats without holes in the hull is a highly desirable state. Documenting all essential processes doesn't have to be a huge undertaking. A lightweight form with a one page description for each process is adequate. There is a natural threshold for how much text people will read so make it easy for them.

Best Practice #4 - *Select technologies that will support your leverage goals.*

Problem Technology choices are often made arbitrarily without adequate regard to the long term consequences. Some companies stay with old technologies far too long because they are comfortable with them. Engineers know how to make the best use of the mature technologies. This path has the innate danger of not embracing new breakthroughs as they emerge. Over several years the company product value may erode and undermine the entire company.

The second, and opposite, danger occurs when companies try to switch technologies too soon or frequently. New technology is inherently unstable and has a questionable value proposition. Many a project has been burned by jumping in too soon. Over time these fledgling technologies stabilize and can offer solutions of real value.

Project leaders (architects and managers) frequently struggle with technology decisions because they don't have a clear set of guidelines that would help them make the proper choices. A bad technology decision can undermine the overall success of the project.

Solution Technology decisions should be made to support several generations of product development. The cost and benefit equation takes a couple of years to develop. This means that if you only base your decisions on the short term you will never invest in new technology at all.

This will put you on a path where your products become uncompetitive within a few years.

Use more established technologies for your mainstream development but keep an eye on the experimental technologies that are emerging in the mainstream. Build a road map for technology, tools, and process that extends for at least three years, and preferably five.

Actively manage each technology transition and allow adequate time for the learning curve. Budget a thousand hours of time per engineer to allow them to become fully proficient with a major new technology. During this time there will be a mix of learning and doing that occurs so don't expect the productivity levels to reach the new goal until the engineers are fully proficient.

Next Steps

- Make an inventory of your key technologies (languages, frameworks, tools, processes).
- Conduct an assessment of each technology and identify possible replacements.
- Create a small investigation into key replacement technologies that hold promise.
- Create a five year road map of your top ten technologies.
- Identify your investment portfolio profile for technology.
- Select two technologies that require active assessment.
- Calculate the learning budget required for any active transition.

Chapter 5 - Architectural Leverage

"Architecture should speak of its time and place, but yearn for timelessness". ~ Frank Gehry

---

High quality architecture is the direct result of applying good design principles and practices to solve the problems at hand. Without a great architecture you have nothing to leverage. If the old solution barely meets the needs then it can't be extended to meet a different set of needs.

If you don't trust the current architecture then don't try to leverage off of it to build something new. You will invest far more time than it is worth, and when you are done, you will have something that is no better than the first version.

Building systems from a fresh perspective, with the knowledge of the old system, is far faster than trying to convert an old system into a new one. A system that costs 1000 hours to build can be rebuilt from scratch in around 300 hours. This is because you already have solutions to 70% of the problems you will face. Design leverage is often far more powerful than code leverage.

The quality of the software architecture directly depends on the strength of the interfaces. Design is all about the interfaces between the components. A reusable architecture has well-encapsulated components and standard interactions between those components.

A poor architecture has very weak components and lots of complex and custom interactions between them. A weak architecture doesn't even meet the needs of its initial use and breaks entirely when you try to extend the business needs to a second generation solution.

Need for Reusable Architecture

Leverage of the design is perhaps the most important kind of reuse that is possible. Reusing key design elements enables other types of reuse throughout the life cycle. However, when designs can't be applied to new development, opportunities to build on other elements are thwarted. When the designs can be recycled it is quite natural to use the product definition, technology, implementation components, and tests to the fullest extent.

Maximizing the design leverage is the best way to optimize the overall leverage of the project. Pay special attention to the way that each generation of project can be built upon the design of its predecessor. High reuse of the design can lead to ten times the productivity and one tenth of the cost. The stakes are high so the effort spent to develop greater skill in this area will be well rewarded.

Design is the Heart of Software Design is the essence of software development. Engineers solve problems by creating solutions that are implementations of design ideas. The code embodies the designs and applies it to the real world. The underlying technology supports the design but doesn't solve the problem directly.

A weak design will lead to a poor implementation and an inappropriate use of technology. However, a strong design can produce a great product that will stand the test of time. A good design is not only critical to the current solution, but can be used as the basis of many other future products. Design leverage lets us get long-term benefits from each design that we create.

Leverage is a test of your design quality. Weak designs can't be effectively reused. The assumptions that were used in creating the original design won't hold for other applications. It is necessary to modify the existing design to accommodate new assumptions. This is where the design quality is revealed. Poor designs are rigid and can't be adapted to new use. Any design that can be easily applied to new situations is strong, by definition.

How Much Do You Leverage? All of us want to achieve a high degree of leverage of our designs. Why build a system that we had no intention of extending? But the reality is far from encouraging. We often get far less leverage than we expect. Here are several viewpoints that are useful to determine the design leverage that you are currently achieving on your projects.

Engineering is done to solve problems. Each project solves a collection of problems. An inventory of solutions can be used to solve the next round of problems that need to be solved. One way to look at your leverage is to evaluate how many new problems are solved by applying existing solutions and how often you need to start from scratch.

Another way to evaluate your leverage is to compare the software budget required to build a new version of your product. This is a good indicator of the overall leverage since leverage is directly tied to the amount of design reuse.

A third way to think about design leverage is to measure the size of your inventory of design tricks. Do you have a substantial and growing list of solutions for common problems? Are these applied to new problems by each team member or are they disregarded? On the typical project, how much new learning must occur? Each of these unique viewpoints can be useful to assemble an accurate picture of your design leverage.

High Leverage is Possible Is it even possible to reuse designs? Many question whether it's practical to build multiple generations of design without starting from scratch. There are several factors that work in our favor. Every company building software works in some kind of a customer domain. This means that we are solving similar problems with each product development. Leverage is possible and required in order to stay competitive. We must learn effective ways to utilize our past engineering work to optimize our next product.

Creating solid designs for our most common problems is an important part of building strong business value. These designs embody understanding of the customer needs. Over time we build a deep knowledge of the domain and the software that serves those customers. The designs demonstrate the subject matter expertise we possess.

Building a Practice of Design Reuse Design leverage is highly desirable, but it is not automatic. It requires learning a set of competencies and applying them every day. It takes discipline to follow through on the practices that lead to success. Initially we need to focus on learning the core competencies that produce good designs. It is critical to set goals for the leverage that we expect to achieve. Modest goals will take a small amount of effort while ambitious goals may require a Herculean effort.

Make an attempt to quantify the amount of design leverage that you are achieving. This will have two benefits, it will help you set better goals and it will help you to justify the improvement efforts. When people begin to see the benefit of the work being done they can get very motivated to support it. A regular review cycle can also make your work more visible to the larger organization.

Evolutionary Design

It is impossible to predict the details of the design that will emerge at the outset of a new project. So instead of putting effort into guessing at the shape of the future you should learn how to build systems that are inherently flexible. Strong design of components can emerge incrementally by applying constant refactoring at the same time that you add features. This overall process of evolutionary design lets you begin development in a direction that will be refined as the system is built out.

Leverage of software architecture results when sound engineering is coupled to a flexible plan. Trying to do too much planning up front will undermine the leverage opportunities. Creating a rigid architecture that doesn't anticipate the types of variations that future system will require will also block leverage. Refactoring is an essential skill that must be used daily to successfully evolve a design without excessive bit rot.

Lightweight Planning Traditionally, software projects relied on lots of up-front planning at the project and architectural levels. This assumes that everything is known at the start of the project. The net result is that the most critical decisions on the project are made at a time of maximum ignorance. The mistakes made early in a project will continue to haunt you throughout the duration of the project. These will ultimately prove to be the costliest mistakes.

Avoid locking in plans that are built on untested assumptions. While there are areas of uncertainty, the focus of planning must be to verify reality in these areas. If technical or market feasibility is a question, find those answers first before creating an elaborate plan that assumes an answer.

In practice, the design and planning can proceed in parallel with the investigation of underlying feasibility issues. But the design work must remain flexible, especially in areas with untested assumptions. Focus on the essential aspects of the design while allowing detailed work to be adjusted later.

This argues for a top-down design. You may know enough now to know that you will need a transaction processing system, but may not know enough to design the nuances of the multi-stage commit and rollback. Save the unknown details for later commitments.

One way to proceed with the design work is to focus on the problems that must be solved first. This creates focus on the essential elements and avoids unnecessary requirements. It also provides a natural means of selecting the most important problems first. The priority of each problem should be proportional to the impact that it has on the project if left unsolved.

Any issue that undermines (or calls into question) the validity of the project overall must be addressed before anything else. The second tier of problems are those that block overall progress in another area. Sometimes, problems must be solved in a certain order and relatively minor issues can stop progress until they are addressed.

Keep your plans lightweight. The plan is only useful if it enables and accelerates real work. No customer will execute your first plan perfectly so the only thing that matters is how your plan will be converted into a running product. Heavy plans slow a project down.

For years we were expected to write a Product Marketing Requirements document, an Engineering Requirements Specification, an Internal Requirements Specification, and a Internal Maintenance Specification. Every project started with an extended period of time to write the hundreds of pages of documentation required. The purpose of this documentation was to reduce risk in the project by making the product development predictable. The rationale was to get the unruly engineers to commit to what they would eventually build and prevent them from wandering aimlessly toward some undesirable product goal.

The tragedy of this approach is that these documents are never even close to how the ultimate product looks. There are so many essential elements that are unknown that any attempt to predict every detail hurts the project. Not only is the effort to write the documents wasted, but the presence of the documents limits the scope of the solution.

Building Flexibility A flexible system is resilient to change. Design leverage is about adapting existing designs for new purposes. This requires that the system be easy to change. Rigid architectures don't change. The initial needs of the design may be easily met but when you start trying to solve a slightly different problem it takes an unreasonable amount of effort to make the changes.

If the design is healthy, a small change is easy and a big change is possible. With a bad architecture, a small change is hard and a big change is impossible. Any change that can be described in six words should be implemented in one line of code.

It should never take 100 lines of code to add one employee record or calculate an order total. If an idea is easy to express it should be easy to implement. If the change is hard then the design is not quite right. Healthy architectures change daily.

We all want flexible designs but flexibility comes from flexing. The design will only honor those constraints that you have forced it to accommodate. If you need a function to take either integers or floating point numbers then you have to design and test those capabilities in. Never assume something works until you have an automatic test that verifies it - testing is the way that you flex a software system. You can't say that your code runs on Mac and Windows until you actually run it on those two operating systems. Your code is not flexible across versions of Python unless you test daily on at least five versions of Python.

If you are building a framework or toolkit that is expected to be used in an environment quite different from the design environment, you must test the full range of operations that might be done by all of your users.

In order to get the benefit of flexibility you must pay the cost of flexing the code. Decide which types of flexibility you need and design a method of guaranteeing that you have achieved that degree of flexibility.

I once worked on a large platform for image manipulation. Our products needed to provide consumers with solutions that would let them edit, organize, enhance, and print their digital photos. The problems that would need to be solved eventually were open-ended - many different requirements would need to be handled by our image manipulation system.

In our case, we built an architectural model built around the concept of an image pipeline. Image data would come into the system and a series of transformations would be applied to it and then a new image would be exported. This problem begs for an architecture that accommodates a variety of image transforms.

The data flow and control flow are set by the architecture but the exact transforms that are needed were supplied as a parameter to the image processing engine. In other words, the flexible dimension is the transform algorithm - everything else is fixed.

Design future flexibility, but stop short of implementing it until you are ready to use it. We can lay a foundation that will meet the future needs of our product but it doesn't need to be fully built out. We should always wait for the business justification before we build features into a product.

As the system matures you will need to build out the full feature set. Addressing secondary features too early in the development process can actually harm the design. Optimize your system architecture for the essential features, then extend the design to handle the secondary features next.

Never build a feature just because you think you might need it someday. Wait for a clear business need then try to get it deployed quickly. In my personal work, I try to stick to the "2 day rule". Don't work on anything that won't be needed in the next couple of days. Of course, as your project scales up you need to be flexible, but the overall goal is to decrease work in progress.

Designing the ability to accommodate future features is important. The cost of design is usually about a tenth of the cost of the full build-out.

It is best to anticipate the future needs within your architecture without incurring the full expense of building, testing, and maintaining the code.

Levels of Design Think about your design in layers of detail. Your design should have a one page description that everyone on your team can recite from memory. This plan should define all of the necessary major input and output elements. It should also break down the top level system blocks that you intend to build. There should be between three and five subsystems that will run your product.

It is important to have a simple to understand block diagram of your system in order to discuss the structure with people both inside and outside of your project. Many engineers are unable to discuss the system without bringing up all of the gory details. These are especially confusing to non-technical people and these discussions can decrease trust.

Within each of these subsystems you will want to use the same approach. Let a one page summary illustrate the core input, output, and functional breakdown. Then extend this into as much detail as you need. Write all of the documentation that you will actually use. There is no time for writing "Write-only Docs".

Every system should be documented with at least three levels of detail. I like to call these:

- Product - overall design to address how external players interact with the whole
- Subsystem - highest level building blocks define all core interfaces
- Module - smaller building blocks that define source code modularity

Below these levels of structure we find the source code for the modules. Each module should also be well-structured with sub-modules and functions. When you get to the modular level you should favor API documentation that is either built into the code or extracted from it.

Paper docs are difficult to maintain so minimize the amount that you write. Documentation that is embedded in the code is maintained as a part of the source code. Better yet, use great naming and let the source code be the definitive reference. After all, your source code will be thoroughly tested daily.

Essential Design Practices Good design involves art and finesse. Less experienced engineers tend to overbuild their systems. They put in far too many features early on and build things that will never be used by a real customer. Since time is spent building extraneous features there isn't enough time to build essentials so they end up cutting corners in critical areas that threaten to undermine the quality.

Focus on the essential elements first. A trick to find out what is truly essential is simply to limit your To Do list to five items. If you only got to implement four features today, which ones would you choose? This technique can be applied repeatedly. Once you finish the next five essential items you can select another five.

Making a list of a hundred features to implement causes you to lose focus. A lot of the power of Scrum comes from the increased focus on the essential. A Sprint is defined for a fixed time period that forces the developers and the product owner to focus on the essential.

Select a few tasks and do them well. Then stop as soon as you have achieved an acceptable result. You can come back later and make it even better but the next action you do should be the most important thing that you can do.

Don't address a secondary issue when there is something of far greater importance that is waiting. You should be constantly asking yourself, "Is this the best way to spend the next hour?" Give special consideration to design issues that may be blocking the activity of others.

Draw a map of dependent activities. It doesn't need to be as complex as a PERT chart. It simply needs to list which design tasks need to be done in order to allow other tasks to stay on track. Then it's good to set dates for expected delivery of that feature.

The graphical view of a schedule isn't nearly as important as establishing the key hand-offs that must take place. Project planning typically involves the task decomposition (with resource planning) and dependency management (with deliverables).

Establish a list of the most critical problems to solve. Then give your team a budget for solving each one. For example, assume that your system needs to authenticate users in order to use your web site. This problem could be solved by a competent engineer in a day. Of course, if you already have a reusable design then you can set a limit of an hour for this task.

Work off a list of problems and make sure that everyone on your team is contributing to solving the most important issues first. A full team can deliver a phenomenal amount of value when you focus the effort. Limiting the time allowed to solve problems is an excellent way to avoid overbuilding a system.

My best development experience was working with a team of five engineers on digital imaging software. We built applications for camera unload, image editing and enhancements, photo printing, scanning, and sending photos. We did the work of over ten people because we only did the essential tasks. A project that has clear priorities is much more likely to be successful because all of the work can be focused in the right direction. A project that doesn't have a clear sense of what is important will struggle from the beginning.

The design must be built according to the business priorities of the project itself and the design effort can be a catalyst in understanding. Sometimes a project lacks clear priorities because the business partners are unsure about the overall business direction. In this case, the software project must develop a software plan that has verification steps for the business ideas as well as the technology. Early prototypes of product ideas can heavily influence the goals of the overall business by answering some questions and asking others.

The Prime Directive - Encapsulation

At the beginning of the computer era programs were written as giant lumps of logic. People quickly figured out that this was a poor way to write software. Encapsulation and abstraction are powerful concepts that allow humans to understand complex software.

Great encapsulation strikes a balance between robust and powerful mental models and simplicity. They are cohesive blocks of logic that remain well-isolated from each other. These components provide building blocks that can be used to composite sophisticated applications. The key is to make it easy to understand the external surface (API) of a component without having to understand its internal details. This is the key to encapsulation.

Mental Model Good designs have at their heart a strong mental model for some type of data. An abstract model defines how the item is represented in the computer and how it is acted upon. Some models are simply calculations while others are persisted in some form of saved state.

A good model must be defined before code is created to implement it. Some engineers rush into writing software before they have adequately thought through the underlying issues of using it.

Our goal is to maximize the leverage. A model encodes a way of thinking about some object within our system. A great model will last for years. The models themselves my end up going through many different implementations over time, and yet, be essentially the same.

A mental model must address the following concerns:

- Data automation
- State transition
- Data flow
- Control flow

Simplicity The most powerful models in our system are the simplest to understand. Everyone knows what a bank account does or how to assemble an order from line items. It should be easy for you to make a list of the top 20 data types that are important in your domain. Here are some examples to get you going:

- Users and login credentials
- Web pages
- Images
- Music files
- Documents
- Accounts
- Transactions

- Graphics (lines, points, shapes)
- Web resources (CSS, JS, HTML)
- GPS coordinates
- Maps
- Aircraft positions and velocity
- Star coordinates

The specific nature of the data types vary greatly with the domain that you work in, but it should be easy to assemble a list of your most important data types. This is an opportunity for huge leverage. Building an arsenal of data models will put you well on the way to building reusable software applications in a fraction of the time.

A simple design should be expressible on a napkin. Build a catalog of reusable designs that meet specific situations. Use the ready-made solutions to address problems as they appear. Parameterize the solutions too maximize the leverage opportunities.

Isolation Changes in one part of the software shouldn't have a ripple effect throughout the rest of the code. If this happens, it becomes impossible to reason through how a change will affect the system as a whole. Over time people will stop making changes out of fear of inadvertently breaking everything, even when changes are desperately needed. This causes the software to rapidly decay and fall into disuse.

Data hiding prevents exposing the details of a module to the modules that use it. Assumptions about the internal details shouldn't affect how the system is used. Otherwise, the two modules are coupled in an inappropriate way. Modules must remain as independent as possible in order to protect leverage. This means that each interface is as narrow as it can be. Each model should have strong cohesion internally and minimum coupling to the outside world.

Building Blocks Our goal is to create a set of building block that can quickly be assembled into applications. This strategy requires that each component be encapsulated so that it is minimally connected to all others. Changes must be contained within the component by minimizing the coupling between the components.

Making components that are useful across a broad range of applications requires extra effort compared to only satisfying the needs of a single use. Don't try to generalize a component too early. Initially you should focus on the specific needs that are in front of you.

With each new usage of a component you can generalize it a little bit. Over time it becomes very general but with this incremental approach you can amortize the cost as you go. Invest around 10% of the effort in generalizing a component each time you need to use it.

Consider the following example. Cindy needs to add a mortgage calculator into her app. To address all of the different type of mortgages she estimates that it will take her about 40 hours to build. However, the needs of the current app can be met with only about 10 hours of effort. She solves the simpler problem and avoids spending 30 hours on a solution that isn't needed yet.

But the next three apps require modifications that the simple solution didn't address. Each app requires another 1-2 hours to enhance the limited solution. In the end, she produces a general purpose tool that is just good enough to meet all of the needs at a fraction of the cost.

Measure the number of problems that you solve using an existing solution and also measure the number of time that you are required to build a solution from scratch. Next, look at the number of times you have reused a particular solution. The average number of reuses should be at least five while the percentage of problems that need original solutions should be less that 20%. These ratios will show you what your true reuse looks like.

Practices for Design Leverage

High quality designs are the product of great engineering practices. There is a short list of core techniques that must be mastered to consistently produce the highest quality designs. Within a team you need someone that thoroughly understands these disciplines and can teach them to the rest of the team. There is no short cut - you must learn and apply these techniques each day.

- Refactoring of legacy code

- Defining interfaces - robust and minimal
- Encapsulation - defining component boundaries
- Building domain languages - allows for testing and automation
- Test-driven development - allows evolution through refactoring

Refactoring The most important design skill to master is code refactoring. This allows you to work with an existing body of code and apply design changes. Evolutionary design requires constant refactoring. If a design can't be modified in-place then it will can't evolve. Therefore, refactoring is an essential skill for working on the design throughout the course of the project.

Let's look at a typical evolution of three product generations to see how this works. The first product is built from scratch with very little forethought to the future. The second application is built by leveraging the first. During this development certain reusable components are identified and encapsulated for later use. The third generation product builds on the second and additional generalizations are made to the code.

The best architectures emerge from successfully refactoring the common functionality out of several different applications. Even though each application was created as if it is the only one, two applications within the same domain will share a lot of common design elements.

A skilled software architect will discover and identify these common elements. Then they can be extracted into a framework that contains the general logic without any of the specific details from the original implementations.

Once the refactored framework is stable, the original applications are converted to use the new framework. This is a fairly mechanical process since the design elements originated from the specific applications themselves.

Now the framework can be used in a third implementation. This code won't contain the application-specific logic that the first two implementations had because it utilizes the generalized logic right from the start.

Inevitably, the third implementation will also require extensions to the general framework. All three applications (and the ones following) will influence the framework by adding additional features. Each new implementation will continue to influence the design, but to a diminishing extent. Eventually, the design of the framework will stabilize.

The end result of this approach is a generalized architecture that is minimal, robust, clean, and well-factored. It can live for many years and service tens or hundreds of projects for decades. Django is a good example of this process. It's a web application framework written in Python that evolved in precisely this way. Several custom apps were developed and each time the common elements were refactored a robust framework began to emerge that developers outside the original team can now use.

This is a high-level description of what happens in an evolutionary design. A rigid design is built at one time for only one purpose but a flexible design is more organic. A design built in a day may last for two years while a design that evolves over many years can live forever. For maximum leverage, look for a design than can morph into something that goes far beyond the original intent.

Many engineers try to create a generalized solution for problems too soon. The reasoning is that they know eventually that certain other problems must be handled and they believe that it will cost extra work to do it later. This is almost always a mistake. It assumes that you know enough about the future problems to anticipate which direction the overall design will take.

Let the general solution emerge over time. This organic growth occurs as features are added and the common elements are refactored into the base framework. Eventually a strong solution emerges that is truly optimized to address specific problems.

The features of the specific app should be built incrementally. The quickest way to build any application is one line at a time. One line of test code - one line of product code. Then pause to remove any duplication that you see. This process of baby steps will let you move extremely quickly.

Refactoring is the most important design skill that you have. It lets you continue the design work throughout the project rather than restrict the design to only the start of the project. I highly recommend that every engineer understand the content of Martin Fowler's book, "Software

Reuse". You'll find that refactoring is a survival skill that prevents bit rot and protects your software investment.

Refactoring is a strategic tool that allows you to incrementally improve your design to match the changing business conditions. No matter how brilliant your designer are they can't see into the future.

Design Patterns The goal of design patterns is to have a standard way to solve common problems. When you encounter a situation that looks like THIS, the best way to handle it is THAT. Build a catalog of these design patterns that work for the situations you are most likely to encounter. Your catalog should also include style guidelines, and programming idioms. A framework is an application that embodies many design patterns. Understanding the principles behind design patterns gives us the ability to build stronger frameworks.

A healthy software organization will have the right balance between autonomy of the individuals and good leadership. A technical leader must oversee the architecture, work on reviewing the patterns, and the performance of the overall system. Creating a public discussion for the top ten design patterns in your organization is one way to encourage pattern reuse throughout your team. Ask people to make contributions to the patterns. As people understand the advantages of various patterns they will naturally propagate best practices throughout the organization.

Interfaces An interface is the connection between two parts of the system. Interfaces and components are the primary design elements. Multiple subsystems (or components) are connected to each other by interfaces.

A well-designed component is tightly connected on the inside and loosely connected to other components. In fact, it is so loosely connected that we can draw an arrow between two components and describe the exact range of interactions that are possible.

Every interface should be tested because, collectively, the interfaces represent the primary interactions throughout the design. Test all of the assumptions that occur on each interface. For example, assume we have an employee model with a name and ID. The interface to employee should contain things like get the name and ID, add an employee, update this employee, and delete this one. These operations are an encapsulation of the idea of "employee" in our system.

Each operation should have test cases that make sure that our employee does what it is supposed to do. Test the employee interface by making employee do everything it is designed to do. Each interface represents some kind of story. Embody the possible operations as test cases or story elements.

Examples:

- An employee is added
- The name and ID are retrieved
- The employee is updated
- An employee is deleted
- A list of all employees is created

Write one or more test cases for every story. You may also choose to create a language that allows you to create your own scenarios. This is a very easy way to automate an interface, allowing you to verify key things about the interactions.

Being able to predict the operations that are present in a given interface provides a significant advantage. The reader can spend more time thinking about the problem they are trying to solve if the calling syntax is familiar. Design interfaces that are patterned after the software other people are comfortable with already.

One example of a standard set of operations is the CRUD API. Implementing an interface to persistent storage begins with defining the implementation for Create, Read, Update, Delete. Additional operations can be added as needed. We can think of a set of standard operations as a design pattern of sort.

Another example of a standard pattern for API definitions is the file interface. This usually is built around a programming model with Open, Read, Write, Close functions. Notice that the operations themselves imply a work flow (involving both control flow and data flow).

As you encounter the same operations repeatedly you should build it into a fully defined design pattern. This lets you address many of the decisions that are inherent in the pattern. Standards leverage learning and help your team avoid floundering when trying to apply the learning. Our key goal is to have our understanding of each problem transferred with the highest possible fidelity.

An engineer can be considered fluent in a language when they no longer have to think about speaking it. Fluency is the result of the interface quality and the familiarity that an engineer has with it. Leveraging common design idioms speeds up the process of becoming fluent. The goal is to minimize the amount of effort required to utilize all of your key interfaces - they should be entirely intuitive.

Some APIs (application programming interfaces) have rich and highly structured vocabulary that can be rightly described as a language. Other interfaces have nothing more than a stack of function calls that take complex parameter arrangements. These interfaces can often benefit from a simple text language that can perform the major functionality of the interface.

Begin by keeping the language simple at the beginning. Let it grow as it needs to, but do not add that burden at the start. Define the key actors in your system and make these nouns. Then define the operations that are appropriate for that type of data. Now you have a simple language for your system.

Build strong data types with standard operations. These should leverage the common understanding and be intuitive to the engineers. Concepts must be simple and obvious. Create simple phrases: "send email me@here.com", "customer update list", "missile launch", "build pdf mybook.pdf". Create a language that matches the customer problem. Minimize the need for modifiers by letting many of the parameters default to reasonable values.

Domain Languages In Martin Fowler's book on Domain Specific Languages, he describes the principles of DSL and how they should be used. I highly recommend this book if you're interested in a deep dive into DSLs. But even if you don't read the book you can benefit from the concept. It's often enough to just define a simple language for talking to your system - it doesn't need the full properties of a DSL.

Implementing a design language gives you endless possibilities for automation and testing. All of your core functionality can be exercised from remote clients. A web app can have a command-line controller and a command-line app can have a web interface. Most of this flexibility comes as a direct result of creating a strong interface language.

Recording commands that flow across the interface can be done just by logging the text of each command as it is executed. This log is useful for humans to study the control and data flow across the interface. A tremendous amount of insight can be gather by watching the actual traffic flow.

It may also be useful to read the log and pass the same commands back across the interface at a later time. This gives the application a record and playback level of functionality which may become central to the daily operations of the business.

Recording commands can be done by capturing the commands that are sent to an interface. Setting reasonable defaults lets you reduce the number of parameters that must be recorded for each command. Try to limit the recorded attributes to two or three. Think about the recording as building a command history that can be used as a script.

Record and playback becomes really useful when you can save and restore state. Capture the snapshot of the system or a portion of it, then create the ability to read that same state directly into the current system. This can now be used as a setup for a testing scenario. Use the following sequence to build a test scenario. This process will net you a robust platform that you can build an entire system upon for regression testing.

- Setup starting state
- Run some automation scripts while capturing commands
- Confirm new state
- Reset the starting state and confirm that changes are removed
- Playback captured commands
- Confirm new state
- Reset the starting state and confirm that changes are removed

Imagine if you could talk to your system. You could instruct it to go through a sequence of events and then confirm the status of the system as a result. This is easy once you have built a simple command language. You can add users, create orders, list accounts, feed elephants, launch missiles, and reset systems. Whatever your system does can be easily scripted. Think of the new opportunities that this opens up for automation or testing. The only remaining work is that of authoring stories.

Start by identifying the characters and actions that are in your stories. Build a language that lets you express the most important scenarios. Pattern your story line to be Turing Complete:

```
Initial State + Transition -> New State
```

Build domain stories that match the things that your system can do and focus on the most important tasks first. Then grow into other areas that you wish to express. Often, you will not get that far because the benefit of building it may not justify the cost. There will be a clear advantage of building your core functionality into a automated system - this will be your primary testing infrastructure.

Best Practice #5 - *Create components with strong encapsulation and standard interfaces.*

Problem Design leverage is central to overall reuse. If a design is preserved it creates opportunities for reusing the technology, the definition, the implementation, the testing, and the planning. The tendency to abandon designs and start over each time is rooted in the mistaken belief that all designs are fragile and only live for a short time.

Many engineers have had bad experiences trying to leverage weak designs in legacy code. They have drawn the erroneous conclusion that design leverage is not really practical. Poor design practices result in weak designs. If the existing code base is riddled with technical debt then design leverage is impossible. But if the code base is sound then the design should be flexible.

If the design can't be modified to meet the new design requirements then there is little opportunity for other types of leverage. If this happens the product cost will be comparable to the original cost of development. This means that the business value of the system is quite low.

Solution Companies are not doomed to reinvest repeatedly in zero leverage software development. Any team can learn the skills required to build well-designed software. Best practices leads to solid designs which support evolutionary design. Strong encapsulation leads to components that are reasonably independent of each other. This in turn makes design leverage possible by isolating the affect of design changes to single components.

Core practices can be learned by any team that desires more leverage. A single engineer within the team can learn and teach others. An entire team using these best practices will increase the degree of leverage within the project and dramatically lower the development cost.

Next Steps

- Measure your current leverage using: % new solutions, budget size, inventory of tricks, required learning
- Set goals for desired design reuse
- Assess your team competency in these practices: ** Encapsulation ** Refactoring ** Interface design ** Languages for your domain ** Record and playback
- Select one key interface to overhaul

Chapter 6 - Code Leverage

"There is nothing so useless as doing efficiently that which should not be done at all".

~ Peter Drucker

The actual activity of writing code is where implementing best practices across an entire team can rapidly multiply your leverage on any project. By focusing on immediate business goals and using standardized tools, your team can rapidly turn out the highest quality software time after time. Now we'll dig into the details of that endeavor.

Tasks within Code Construction

Earlier we discussed the different development tasks that must be balanced throughout the project. These tasks make up the essential work of implementation. Coding requires four different types of fundamental activities. Each type of task is built on a unique frame of reference. As code is built these viewpoints must be expressed and held in balance throughout the development cycle.

- Test - verify functionality
- Fix - repair defects and errors
- Extend - new features
- Improve - structure and performance

Development Tasks Must be Balanced

If any one of these tasks is neglected you will build up technical debt. This can have disastrous results on a project unless it is corrected. Each of these types of work has some key best practices associated with it. Maximum leverage is achieved by solving the key problems in each area. Be careful to budget your time to service each of these area with an equal amount of effort.

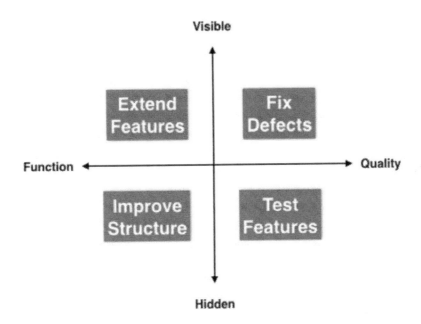

Figure 9: Development Task Types

Each of these core perspectives has its own form of technical debt which results from neglecting the required work. Each task also has its own best practices that prevents the accumulation of debt. This chapter will explore each task and show you how to overcome the related challenges.

Test - Verify Functionality

The speed of development is directly related to the speed of your testing. Anyone can create a large amount of code and fail to integrate it into the overall project. Testing is central to the coding effort - without adequate testing your team will spend a large percentage of their time trying to debug mystery problems that come and go.

The subject of testing is so critical to development that I've dedicated the next chapter to it. In this chapter we will focus on how testing affects the coding process itself. We will focus on the integration of coding and testing and leave the mechanics of testing for later.

Before a new feature is added you should think about how that feature will be tested. Once a test has been constructed for a desired feature it is easy to implement the feature properly. More importantly, it gives you confidence that features that were previously working properly are still working and it will save countless hours of debugging things that were accidentally broken. Extensive debugging is a sign of poor test coverage.

Begin with Tests Before you start working in any area of the code you need to think about how this can be tested. How will you know three months from now that the feature isn't broken? Writing a test lets you know when the feature is complete and that same test will prevent the feature from being broken without you knowing about it immediately.

Write requirements as tests. Use the tests to specify what the product must do. Be specific about the starting and ending assumptions - the test should reflect what you assume to be true. If you write a line of product code then it should be tested with a line of test code. Ultimately, you will create as much test code as product code. This is one way to get quick feedback about the quality of your coverage. If only 10% of your code is tests then you aren't getting adequate coverage.

Defects discovered throughout development are another piece of feedback - it is a clue that a required test is missing. Create a test for each defect that you discover in your product, thereby filling the gap to prevent this defect in the future. Prevent defects from becoming repeat offenders; recidivism is another sign of poor test coverage.

Limit to One Line of Code The best test is a simple one-line assertion. Verify all of the assumptions that you believe should be true. If your assumption doesn't hold then learn from it. A test should be simple to write and simple to correct. If you believe that there should never be less than thirty customers in the database or 100,000 files on the file system then express that as a single line of code.

Test an assumption with a single assertion, you don't need to get fancy. Just pick a test case that verifies one situation that you know must work. Each line of product code has a test, so the tests should be short and sweet. Test concrete examples rather that representing a large and complex set of interactions. Our goal is to catch stuff that breaks, not to present a formal proof of correctness. If your tests become complex, throw them away and replace then with stacks of simple assertions.

Group related tests into a test suite. Use the module level to encapsulate tests that match the product modules. Use functions to group things that are testing product functions. This way the test structure roughly mirrors the product structure. If your product has 30,000 lines of product code you will also have another 30,000 lines of test code. Make sure that your test code is well organized with clear correlation to your product, otherwise it code maintenance becomes difficult.

Testing Framework Over time, you will gather all of the assumptions in executable form. In ten seconds you can verify everything that you have ever assumed about the entire system. If you run these tests hourly then you guarantee that you know within the hour when something unusual happens.

To make writing and running your tests efficient you will want to set up a simple framework. It could be as simple as a test program with a thousand assert statements. However, you will almost certainly want to have a more convenient way to view and respond to the test failures. In the next chapter we will lay out the specs for a system that will provide an easy way to review and manage test results.

It is important that a single command will run all of the tests for you quickly. Experience shows that if tests are difficult to run, they will not be run at all. Spend the time to optimize the speed of your testing. Build your own design rules about how long it should take to write or fix a test. Set a time limit on how fast you need to get test results and automate the running of tests with a system scheduler, such as Cron.

Make sure that your tests are running at least hourly. You can run complex tests (that require a long time to execute) less frequently. But these should also be fully automated and run daily or weekly. If the tests are not automated, then it is unlikely that they will ever be run more than three times.

Tests are critical to your success. An adequate test battery can verify a thousand assertions in a few seconds. You can be confident at all times that your code is still working properly. This is extremely important when you are leveraging the code into a new environment.

All Tests - Every Five Minutes Make sure that your tests run fast. By default you should force tests to yield and answer in a second. As test times increase, developers will stop running them and the primary value of quick testing can be easily lost. If you feel like you need to have long running tests then cache the results. Save the results for a hundred times as long as it took to calculate. This simple trick will allow you to run shorter tests all the time and still run long tests occasionally.

While you are coding you should have a super-quick test that just tries the one thing that you are working on. This should be assigned to a button press or a command like 'x'. This will let you verify one step in less than a second. Once the small thing works you can test the entire system in another ten seconds. Run the small test several times a minute and then the full tests every few minutes.

Each command that you execute has the potential to be a good test. A test is nothing more than a coded assertion so any script can be a valid test. If the script dies in the middle or fails to produce the expected results then it is a failed test. A command can be repeated later with minimal effort. A test runner can execute all tests, which makes retesting effortless. I often use the "history" command on Linux to review recently executed commands to look for possible new tests.

Fix - Repair Defects and Errors

The second major task required during coding is to fix all of the problems that are introduced. Nothing is perfect on the first attempt to implement it. These problems must be resolved early or they will undermine the fundamental integrity of the system.

Bugs in a system will breed like roaches so don't fail to act as you encounter bugs. Build an urgency for defect-fixing directly into your coding process itself. Every command that you execute will either produce an expected result or something that you didn't anticipate. When you are surprised, stop and investigate. This may be the only warning that you get that something is wrong.

Logging every issue that appears is a good idea, but far too often this is the only action taken. Why not attempt to solve the problem immediately? If you make it your habit to solve every problem as soon as you see it then it quickly becomes difficult to find unsolved problems to work on. This frees you up to build more tests and more features. Remember, one defect can be hiding a far worse one. Fix the first and the second will surface.

Try, Fail, Recover When you execute any command you assume that something will happen. A surprise means that your assumption was wrong. Stop and think before you move on. Your next action should be to verify the assumption about what just happened. Don't neglect to acknowledge the surprise, or fail to learn from it.

Trying to write too much code before testing it is a serious mistake. Even when you know it is wrong you can still be pulled into an attempt to write a full block of code. Resist this urge and write one line of code at a time. Then write a single line of test code. This method may be counter-intuitive but you can move much faster overall.

You should expect to clear about 100 lines of code per hour when working in this fashion. If you ever find yourself debugging and trying repeatedly to get things working, then it is a process failure. Stop, backtrack, and start taking baby steps again.

Each step you take should get an expected result. If you are surprised, don't move forward until you can explain why. Fix an unexpected problem with one line of code if possible. Try to keep the edit/test loop going at least once a minute.

Adjust Scope Based on Each Trial Don't make repeated attempts to solve a problem. Instead, change the context. After two failures, isolate one part that you suspect may be causing the failure. Zoom in on the problem by changing the execution context. Keep executing less and less code until you have a test case that executes one line of code and demonstrates the problem.

Consider the example of importing a file with employees in it. The error message says 'Import Failed on record 24'. The natural impulse is to start changing random things and trying to import the file over and over again. Instead, try to feed line 24 directly into the import record function. By doing this, we have moved from a "read table" context to a "read record" context. With this approach you're more likely to quickly find something useful like the issue is with the name lookup during the import.

Keep zooming in until you find and fix the problem. Then zoom out to each level that previously failed. Eventually, all of the tests that were failing will pass. Save the test cases in your test suite and never again will this problem happen without you knowing instantly.

Here are some simple rules for your development loop:

- Reduce scope of context after each failure
- Enlarge scope of test after each success
- Don't attempt multiple fixes in the same context
- Minimize the code in each iteration
- Enlarge the scope all the way up to the full product test
- Debugging is a red flag indicator of poor testing
- Run the loop each minute
- Sprint and rest
- Measure your output

Goal Stack It is often helpful to keep a goal stack while working on a new feature or resolving a defect. Things come up and require your immediate attention, but once resolved you want to switch your attention back to the previous goal. This is a lot like how subroutines work.

Start with the biggest goal. Then each new issue causes you to push the stack. Resolving an issue pops the stack and the stack itself provides a reminder of how you got here.

Example Goal Stack:

- Build authentication system for users <- highest level goal
- Build login user story
- Create http://xxx.com/login
- Build Django login view
- Fix CSRF issue with page
- Enable Django middleware to allow CSRF handling
- Research issue on Stack Overflow <- lowest level goal

Start by selecting a single product feature to work on. Avoid all distractions until this feature is created, tested, integrated, and committed to version control. Try to break the larger feature into smaller parts, then push the first part on the stack. Focus on just the current task before you add a bunch of other tasks to the list.

Push new problems on the stack as they interrupt the work in progress. Try not to let the stack get too deep since this represents open mind share that requires energy to manage. When the current problem is solved then pop the stack. An empty stack means that you can select the next feature. This process will drive full closure on the work in progress before starting new work.

Seek Closure Coding is a complex task and there are a lot of minor issues that must be resolved in order to get closure. Multitasking can be a major source of defects because of a fundamental limitation of the human brain. Trying to focus on more than one thing at a time rapidly decreases the effectiveness of cognitive ability. Therefore, concentrate on a single area until it is fully implemented.

Balance the different types of work required for a given feature. Make sure that Test, Fix, Extend, and Improve goals are all matched. Don't neglect fixing issues or writing tests. Solve these issues while your brain is fully loaded with the details of the problems. The best time to refactor any given area is immediately after adding new functionality. Improve structure before you leave the area. Resolve work in progress now so there is no lingering doubt about its status.

Extend - New Features

The third major task during coding is to add new features to the system. This should start by creating a quick test that tells us whether the feature is working properly. Most features should be implemented in several levels of detail so that complex features are composed of smaller ones.

Extending the product will involve implementing each of the sub-features making up the larger ones. Just as the high-level features should start with a test, so should each of the detailed features. Think about the features as being defined by an outline of requirements. Each requirement should have a test case that defines when the requirement is fully complete.

The coding proceeds from the most detailed features to the highest-level features in the system. After everything is complete then all the tests will pass and the code can be committed to version control. This development loop allows you to quickly iterate over hundreds of specific features rapidly to reach the end goal.

Personal Automation We all face many problems that could be fully automated. As programmers, we have an array of tools at our disposal that could be easily used to script any repeatable task. If you have to repeat a series of steps, create a script to do the task for you. Automation is a great personal tool for removing any routine tasks from your daily work flow.

Some tasks are simple enough that we never think of automating them. But even simple tasks carry a burden of memory and judgment that may either be forgotten or misapplied when needed. Even when it is done correctly it still creates a mental burden. Consider the following shell script:

```
copy_files:
    rsync -auv My_System:Projects/Hammer/ ~/Projects/Hammer
```

This script copies files that are newer from one directory on another system to my system recursively. Note the density of the encoded knowledge and design trade-offs. Sure, you could type it in each time you needed it... but why? Running 'copy_files' gets you the result that you want without thinking about how this is accomplished.

Generating scripts gets you to a phenomenal level of code leverage. I have been investing heavily in automation over the last few years and now I have my own command language that has a vocabulary of well over 1,000 scripts. Almost anything that needs to be done can be accomplished with two words on the command line.

You may not need to go quite that far in order to achieve impressive results. Automation will pay huge dividends - one hour spent automating will typically return ten hours within a month.

Code Generation We run into many cases where we can use a parameterized template to create unique code. All web frameworks and IDEs have built-in tools that can be used to generate code using a template. It is also quite easy to build your own code generators in Python, Ruby, Shell, or PERL. If your problem has some boiler-plate code in it then avoid generating this code by hand.

A scripting language can let you capture the commands that you enter manually on the command line and turn these into a script. For example, if you are working to set up a computer you may execute many commands before you find the winning recipe. Once you find it the code may be reused as a script.

```
history > setup
```

Set your context so that it is easy to create, edit, and execute commands. Make sure that the commands are created in a place where they can be automatically executed without ceremony. Imagine having a script that sets up a new command for you from the command line.

```
cmd-add list 'List my files' 'ls $xxx'
```

If the code that is created has repeating structures within it, you may find it useful to write a simple script that creates the repeating code for you. These scripts can often be plugged directly into your text editor to extend your ability to write code. Think like a programmer when you are writing code - if you can simplify the task of writing by creating a short script that writes the code for you, do it!

The final type of code generation to consider is creating programs to transform existing code. These are often based on regular expressions. A simple program can save work because it can be filed away and resurrected or created when needed. These programs may start out very trivial but often evolve quickly into something that is vital to the project. They frequently end up as embedded tools in a larger automated effort. Consider the following simple example.

Fixer.py:

```
from sys import stdin
text = stdin.read()
text = text.replace('this','that')
text = text.replace('-','_')
print text
```

Static Analysis Tools Source code is encoded knowledge - it expresses a solution to a problem in some type of a programming language. There are tools that can analyze the language itself to give us insight into the structure of our code. We can learn a lot from analyzing the code that we produce by catching errors and revealing inefficiencies in our coding.

Style checkers can be used to look for superficial problems within our code. We can look for formatting issues and simple design rules violations. The Lint family of tools are unique to each programming language environment. These do a thorough analysis of the correct usage of the language and they can detect everything from uninitialized variables to bad return values.

You should be using these tools and building them into your project infrastructure so that they run automatically. Learn how to suppress the errors that you aren't interested in. It can be a pain to set these tools up the first time, but their presence can prevent a lot of errors over the long term.

The last essential set of tools will help you monitor the complexity of your code over time. Later in this chapter we will discuss how you can best manage your source code complexity. Each of these individual tools will give you good results and combining them together will create a great toolkit for productivity.

Editing Code There is a seemingly endless debate about using a simple editor or an IDE (integrated development environment). I have been in both camps so I'll just say pick the tool that helps you get the job done quickest. In fact, it can be very beneficial to switch between them. If you are an IDE guru, consider switching to Sublime, and the reverse.

An IDE isolates you from technical details, while a programmer's editor makes it easy to build your own tools. Both of these tool types offer advantages and both of them come with limitations. If you are skilled at both types of tools then you will have a sense for when one will be more effective than another.

The editing environment must support a productive work flow. This requires a few key features. Your work flow must let you have an extremely fast development loop. You need to be able to edit the code and run the tests in less than 3 seconds. This means that the source code and executable program must be visible at the same time.

Other features that you need for a modern editor are color syntax and command completion. These tools act as an extension of your brain during coding and save a significant amount of time. The final useful feature is refactoring tools like the ones built into the JetBrains products. These will help save a great deal of time by doing the grunt work of editing during refactoring.

The most essential features for editing:

- Fast Edit/Test loop (two windows works fine)
- Add a snippet (using a template with parameters)
- Color syntax highlighting
- Command completion and API lookup
- Jump to definition and usage
- Plug in built tools for transforming code
- Refactoring tools
- Static analysis
- Integrated debugger

Rapid Prototyping Rapid prototyping is valuable because a new system is often far easier to create than modifying an existing one. After all, the new system has minimal complexity while the existing product may have hundreds of constraints. A new problem can be solved quickly by creating a new environment where there is a simple context available.

Once a problem is fully solved then the source code can be moved from the simple test application into the product. Consider using the following process to build each new feature:

- Isolate - build a stand-alone app that just contains the desired feature
- Invent - create detailed feature set that is desired
- Integrate - combine your new code with the existing product logic
- Regression tests - save all of the test logic so that it can be run hourly

A full-featured product will emerge much quicker with a rapid prototype process. Each time that you have an idea to try out you can generate a stand- alone app that lets you work on the idea all by itself. Think of it as a work bench where you can experiment with new product ideas - this doesn't need to be the same as the product context.

You should be able to create an entire prototype in less than a day. This gives you the ability to try out a new application idea with a very small investment. You should be able to go from the raw business idea to a running application that embodies the core functionality very quickly.

A work bench application lets you experiment with specific algorithms in isolation. Your product domain will dictate the specific data types that your application should support. For example, if your technology area is imaging you will need a work bench for trying out the effect of different image transforms.

Our goal is not simply to get to the first product quickly, but to create tools that will let us produce a family of products quickly. Solve the meta-problems in order to encourage the largest possible leverage. Think about code prototyping capabilities that will benefit you now and two years from now. Then build the tools that you are missing today.

Improve - Structure and Performance

The final task of coding is to make improvements to the structure and performance of the system. Many engineers are happy once the features work properly and they fail to work on the underlying structure. Over time, each feature introduces an unpaid technical debt. The only way to prevent this from happening is to commit about a quarter of the coding time to refactoring the code.

Refactoring is often viewed as an optional phase and just extra unnecessary work. However, improving the structure of the code is as necessary as building the product features. If the structure decays with each edit the system will very quickly be rendered useless. No software can be reused when it is riddled with structural flaws and duplicated code. Structural improvement is mandatory if you intend to have any leverage at all. Each time you add a feature make some small structural improvement to the immediate area. This guarantees that the more you touch it the better the code will get.

Build Reusable Components Creating reusable code isn't difficult because there are a few simple principles that govern the flexibility of code. Code that is well-encapsulated can be used in environments that differ drastically from the original context. Code that isn't connected to all of its surroundings won't break when the context changes. This is the most important principle for code reuse.

The next principle that enables us to reuse code is to have a clear way to extend the functionality beyond its first usage. You must be able to flex the code in a new direction that doesn't risk breaking everything that was already working. Build the proper amount of flexibility into the design because some applications have a higher need for flexibility than others. Code that can't be tested easily will never be flexible.

Components must be testable as stand-alone code. Testing components only in the primary application that they are built for is completely inadequate because there are far too many special conditions that must be exercised to feel confident about the testing. Automatic tests are necessary for each software component in the system - you must be able to validate each part in isolation.

Every system has a few key interfaces that do most of the work. Consider building a language to interact with the key elements of your API. This lets you build scenarios to create data in your system, capture key traffic across the interface, and playback transactions from earlier recordings. These remote control capabilities can greatly extend the reuse of your software.

Consider building a simple program that exercises a particular interface and use it for many different purposes. Interacting with a component directly, rather than indirectly through the rest of the system, can be very enlightening and is useful in a wide variety of debugging and analysis scenarios.

Refactoring Refactoring is the primary skill required to improve software. Without this ability you are doomed to creating software that can't be maintained. Project managers frequently don't understand the value of refactoring, but it is a vital part of any development. Refactoring is what keeps your product healthy over the long term.

The power of refactoring is built on having a comprehensive set of unit tests. It is impossible to do large-scale refactoring without tests. You may be able to get the code running properly at the start, but you will never have the confidence that all of the code is working in every case. After a disaster strikes your project the decision makers may come to the false conclusion that refactoring was the cause of the failure. This can cause you to lose support for any future refactoring operations and without refactoring your system will quickly become both rigid and fragile.

Refactoring is the key to flexibility. By continuously making small adjustments to your system structure, the system can remain healthy indefinitely. If you stop making improvements to your structure then the design is etched in stone and can't adapt to the changing world. It is imperative that your design remains flexible. Your product won't meet its fundamental goals unless it can be maintained over time.

The end state of most refactoring work is to eliminate duplication. This comes in many forms in the software. Learn to recognize the smell of duplication within the code:

- Lines of code that are repeated
- Algorithms that are written more than once
- Similar logic patterns that should be consolidated into a function
- Poor encapsulation that will cause future duplication

Build your own catalog of refactorings. Here are the most useful refactorings in my catalog:

- Extract code into a function
- Move function to another file
- Rename function
- Rename variable
- Pass state as a parameter

Complexity Systems naturally grow more complex over time. Because complexity lowers quality, it makes the system more costly to maintain. Complexity must be managed throughout the development cycle in order to control costs and preserve quality. Yet, in most organizations there is a lack of appreciation for the impact that code complexity can have on a project.

Complexity is the key driver of software quality. If your system is complex it will be far more costly to build and maintain. The cost of a system grows exponentially with the complexity. If you want to control something, you need to be able to measure it. For this reason, I believe that it is vital to have a complexity measure that can be computed within 5 seconds.

We need to start by understanding what causes complexity in the first place so we can use a tool to measure the complexity of the code. The complexity measures will point out areas of the code that need to be redesigned. Finally, we can develop simple rules that all engineers can use to constantly decrease complexity.

There has been a great deal of good computer science research on code complexity. McCabe and Halstead have developed well-calibrated methods of computing complexity numbers. In practice, I think this degree of rigor is seldom needed. Instead, consider simple metrics that can be computed in real time by a script.

Complexity comes from all of the combinations of all of the bits of logic that occur in our systems. Complexity is driven by two main forces - the size of the system and the connections between different parts of the system.

I like simple tools that allow me to reason through a system. Even if they don't produce "accurate" results, they will produce useful results. Over time, we can learn to interpret the results within a given context. We can use this principle to build our own custom measurement for complexity. In **Appendix A** you will find step by step instructions for how to build a code complexity measurement tool that is customized to your application. This can be altered to reflect your own perception of what complexity means in your world.

The major advantage of building your own complexity measurement tool is that you can start simple and evolve your own understanding of what makes your code complex. Your opinions about complexity can be directly reflected in your measurement tool. Over time you can steer the design to reflect the attributes that are most valuable to you.

The complexity metrics will help you understand the source code better. Each module is listed with a corresponding measure of its complexity so the tool points out modules that are far more complex than their neighbors. The exact value of the measurements are less important than the relative ranking of the modules.

Run the tool repeatedly and develop simple rules. When a module gets too complex, restructure it. Create a threshold that triggers an automatic review or rewrite. These metrics serve as an analysis tool to increase your understanding of the underlying dynamics of your code. Regular usage will make you adept at finding and fixing hot spots. Over time your complexity will decrease simply because you are paying attention to it.

The exact values produced by the metrics mean nothing, but by quantifying the complexity, it creates a baseline of knowledge. Build this understanding over time and use it as a key method of steering the project.

Remember that cost scales with complexity - decrease the complexity in order to decrease the cost. This simple tool will point you to the biggest areas for improvement. Then, focus all of your energy on the most complex parts rather than trying to simplify everything.

```
Complexity => Risk => Cost
```

Make this incremental simplification a normal part of every day. Replacing a complex thing with a simple thing will make everything better. Imagine the effect of 1,000 minor improvement to your project. Each lowers the risk to your project and collectively they can easily yield a profound reduction in cost.

Version Control Version control tools are the most under-appreciated tools in programming. Embracing the strategic value of version control can dramatically accelerate your leverage. Modern tools like Git, Mercurial, and Subversion are at the pinnacle of a long string of tools that began in the 1970s. Make sure that you are using one of these tools and begin to build your entire development stack around the version control tool. I will refer to 'git' throughout the remainder of this book but any of the modern tools will suffice.

Source code is anything that is a fundamental expression of the design. The source code is combined with the tool chain to produce the deployed product. In essence, the tool chain itself is a part of the source code. However, the tools are not usually stored under version control since the exact versions of the tools can be installed from scratch when needed.

Source code also includes all of the operations information, such as server configurations and update scripts. Anything that is needed to produce the running software must be either versioned or readily installable from another source. It is important to have automatic processes to setup and configure all of the different types of servers with the required tools.

Like everything else, we trust but verify. Never believe that you can build the product from scratch unless you do it weekly. Your entire build process must be fully automated and run periodically. Tear down the world and reconstruct it with nothing more than the git repository and required hardware.

Version control is the one point of integration - it is the primary method for developers to coordinate with each other. The developers should have clear rules about how to branch and merge. Divergence is the enemy so you should try to integrate code hourly. At the same time, you also want to clearly identify when it is appropriate to work on a development, integration, or staging branch. Proper rules for how to use version control is too important to be left up to chance - it won't just evolve on its own. Develop your own custom set of practices and train each engineer on the correct techniques.

Best Practice #6 - *Use a balanced approach to development resulting in minimum code complexity.*

Problem Developers will often be sensitive to one aspect of the software while neglecting others. This can easily create an illusion of rapid progress. However, late in the life cycle the project will start to fall apart. This is a symptom of problems with the software process from the very beginning of the project that only become apparent when it is the most difficult to correct.

Poor process results in lopsided development and is responsible for creating pockets of complexity within the system that make the overall product hard to understand. Most long-term maintenance problems can be tied to complex designs that have been implemented by complex code. As time progresses the complexity of the code will threaten its very viability and cause an early replacement of the entire system.

Most teams don't have a clear picture for the code complexity because it isn't being measured or managed. Yet, complexity is at the heart of many quality problems. Improving structure and performance require that code complexity is readily visible.

Solution Let complexity measurement drive development. Complexity is a problem that can easily be solved. The first step is to have a tool that can reveal the most complex areas of any code so the problem areas can be simplified. Once a complexity measure is in place it prevents new code from becoming complex without anyone being made aware of it.

Legacy code that was written at an earlier time can be simplified as well. A balanced approach to software development requires that developers pursue the clean up of structural issues and fixing defects at the same time they are building the system. In the case of legacy code, a certain amount of remedial work is required to create tests and fix structural problems before any new functionality can be added. This makes working with legacy code more expensive than starting a new app from scratch.

Whether you are building from scratch or working with legacy code, you must be able to refactor. This requires the development practices to support it. If you can't refactor and evolve your code it is only a matter of time before complexity takes over your system and your only choice will be to replace it.

Next Steps

- Conduct a self assessment of time spent on: Test, Fix, Extend, Improve
- Select the most critical area for process improvement
- Measure your test code (in lines) and compare it to your product code
- Measure and assess the number of open defects on your project
- Try out either the dev loop or goal stack idea to accelerate coding
- Implement a complexity metric for your source code

Chapter 7 - Test Leverage

> If you don't like testing your product, most likely your customers won't like to test it either.
>
> ~ *Anonymous*

Testing - Traditional or Practical

There appears to be a gap between the traditional ideas of testing and quality assurance and the very specific methods and techniques taught by the advocates of Test-Driven Development. The classic discussions focus on planning and building large documents to capture the Test Plan, Risk Analysis, Test Strategy, Test Tools, Test Cases and Defect Tracking. The primary artifacts generated are documents and the core testing activities aren't automated. Large numbers of people are hired to interact with the system.

The other main school of thought is based in the principles of Agile Software Development. The key ideas here deal with automated testing and unit test frameworks. These are used throughout the implementation phase and serve to ensure the quality long after the implementation is complete. I believe that these concepts are the foundation that every testing effort should build on.

Test-Driven Development, in the form most often taught, assumes that you begin using the correct testing approach at the beginning of the project. Tests are built as the product is built. The two parallel infrastructures validate each other by confirming all of the embedded assumptions. Development with two parallel structures that are mutually reliant on one another is difficult to achieve after the fact.

What if you are given a million lines of legacy code? Is it really practical to tell your stakeholders that your first job is to produce a million lines of test code so that you can do reliable refactoring? Of course not! We need a method to apply the principles of TDD to existing systems.

This chapter attempts to fill the gap between the ideals of TDD and the practical realities of legacy software. We will remain true to the ideals of rigorous testing while making it easy to add and maintain tests that utilize existing parts of the system. In order to achieve the maximum leverage, we need test strategies that can be used effectively for both greenfield applications and legacy code.

Expected Results

Traditional testing techniques have many problems that end up compromising the overall test effort. In the previous chapter we introduced a new style of testing that works well for test-driven development. We will continue to build upon these techniques to show a broad range of testing strategies, tactics, and tools.

The techniques taught here are useful over a broad range of applications and software types. We will demonstrate how to very quickly create and repair tests and assemble them into a huge inventory of test cases. Small test shims can also be added to any type of program to exercise the built-in product functionality for the purpose of testing. We will examine how to construct test cases for different types of situations. Finally, we will wrap up with a look at building an effective testing strategy.

Every Test has Output Our idea of testing is based on every test passing out some type of output. Some tests will pass out text that represents errors that occur. These tests are typically silent if everything works correctly. Other tests may pass out lots of data that is the same each time the test is run.

When a test is run for the first time the output is captured for later. When the test is rerun, the same output is expected. Each time the test is run the actual results are compared to the expected output. This ends up creating a very simple contract for each test.

- run the script
- capture the output
- should be the expected results

Unexpected Results The Unix diff command is used to detect the unexpected results. These differences are then shown to the tester as a test failure. The assumption is that the test should produce the same output each time it is run. If not, there is a mystery to investigate.

In real life the output of many tests contains results that vary each time the command is run. This means that without additional work put into the test, it will fail every time. In practice each test starts off noisy and then it is modified by filtering the output to quiet it down. For example, consider the following test 'ls -l', which produces the following output.

```
total 72
drwxr-xr-x@  26 markseaman  staff    884 Apr  6 15:48 Code
drwx--@      19 markseaman  staff    646 Jul 11 09:18 Dropbox
drwxr-xr-x@   3 markseaman  staff    102 Feb 21 15:16 Money
drwxr-xr-x@   9 markseaman  staff    306 Apr  7 11:33 MyBook
drwxr-xr-x@   2 markseaman  staff     68 Apr 10 07:27 Notebook
drwxrwxr-x    5 markseaman  staff    170 Jun  7 09:00 logs
drwxr-xr-x  153 markseaman  staff   5202 Jun 23 08:45 test
```

This works perfectly if you want to watch any files that change, but it will fail repeatedly if the files are changing and you don't care. In that case you would switch to 'ls' to ignore the things you don't care about. Here are several ways to silence noisy tests:

- Ignore all output - This will mute the test except for errors.
- Look for special patterns only - Mute everything else.
- Remove problematic text from output - Use "grep -v" to remove stuff.
- Count lines only - Look only at size of the output (min, max).
- App logic that knows the content of the output - Pattern analysis.

In general, you should try to use the simplest thing that will work. To be effective at Diff Testing you need to be able to resolve a test problem in less than a minute. The filtering techniques listed above are in the order that yields the simplest solution first.

Approve Results Any test that has unexpected results for its output is considered as a test failure. But these results don't always correspond to an actual system problem. They simply inform you that there is a mystery. Frequently you can look at the unexpected result and immediately understand why it was reported. This can be approved so that next time it is the required answer.

The system is deliberately constructed to bring your attention to things that are changing. Often this is a confirmation that the last action you took had the desired effect. Once you recognize what the test is telling you it can be approved so that this answer will be expected on all subsequent test runs.

For example, all of the text files used to write this book are processed with the following command ('wc -w *') to count the number of words. Adding a new chapter or editing the chapter text will make this test pop. The result can be approved with the command, 'tst like book-words'. Now the result that I just got is the new required answer. This makes accepting new results a trivial operation.

In real life, you build hundreds of tests which are run all of the time. Almost everything that you do changes something somewhere somehow. Your tests find these things and bring them to your attention. Then you look at all of the results and approve them one at a time - you can see quickly what failed and why. If many tests fail and you scan the results, you may want to approve everything in one shot with:

```
tst accept
```

This is exactly the same as visiting each test and approving it. This gives you control over what is approved without having to do the tedious work of approving each test result individually.

Dealing with Failure Not every output that comes from a test is correct. Sometimes the unexpected result is a symptom of a real system failure. The value of our approach is that you can watch a thousand things and the system tells you to pay attention to the ten unexpected results. Diff tests focus my attention to the areas that may have problems.

A test failure may be resolved in several ways:

- The failure is intermittent and goes away with the next test

- The test is noisy and its output should be filtered more
- The test is bad and is reporting a false positive
- The unexpected result should be approved
- The product is broken and should be fixed

The nature of the output makes it very easy to troubleshoot failing tests. It is usually possible to fix ten failing tests in less than one minute. In fact, many times you can spin through the results (using 'tst results') and then approve all results with 'tst accept'.

The real failures are fixed by either improving the product code or the test code depending on which is really in error. In practice about half of the errors are in the product code. This is consistent with the idea that half of all the code is product code.

Speed Goals

The most important goal of a test system is that it is easy to create new tests, run all tests, and fix failing tests. If these tasks are difficult, testing will be abandoned. Trying to implement TDD in a project and not succeeding is far worse then never trying at all.

I have been practicing unit testing for 25 years. The most common reaction I get when trying to get an organization to use the techniques, is "Oh, we tried testing once and it didn't work for us". As a result, these organization have abandoned testing altogether in favor of lots of people using the product and logging defects. Another recurring fallacy is the belief that developers can do their own testing. Most engineers tend to be poor testers unless you can redefine testing to be a development task.

The reason that most attempts at integrating automatic testing into the software development process fails is that it is too hard to maintain the tests. If it were easy to create, run, and approve tests, everyone would do it. The techniques proposed here will let you do just that.

In 15 seconds you should be able to ...

- execute all tests
- create a new test

- approve the answer you just got for a test
- modify the test case

If any of these tasks take longer than a minute then you are doing something wrong. Speed and ease are far more important than closing one small loop hole. Your goal should be to create a thousand simple tests over time. Each test should be roughly one line of code. If it is that easy then tests will create themselves. Testing will become an automatic task and engineers won't even be aware that they are doing it.

Time to Create a Test The first performance goal is the time that it takes to create a simple test. A command lets you type a single line of code to create a new test with a simple function. Here are a couple of examples:

```
tst add list-files 'List the files in my directory' 'ls'

tst add code-length 'Count lines in code' 'wc -l xxx.cpp'
```

You should be able to quickly perform any of the core actions within 15 seconds. In an hour you should be able to generate close to 100 tests for a system that you already know quite well. This is perfect for when you are trying to implement numerous tests for a legacy system. A competent engineer should be able to wrap an entire system in tests in about a day.

Now, let's look at how all this works. Tests are just shell scripts that produce output. We could get fancier than that but we don't need to now. When a test is executed we capture the output including any errors that were produced.

Time to Run Tests Now these new commands are part of our arsenal and can be run with the 'tst' command. The status can be reviewed with the 'tst results' command and all tests will be run automatically. Every test must run without requiring any input to be provided. A complete test battery should run in less than 10 seconds.

There are two strategies for preventing a delay for tests that require a long time to run. The first is to group all of the long running tests into a separate test suite. While doing development, the fast tests are run frequently and the slow tests can be run automatically on a daily or even weekly basis to test complex system operations. When we discuss systems operations maintenance we will revisit this idea.

The second strategy is to cache long running test results. Diff tests have a built-in caching mechanism that does this for me. The execution of each test is timed and the result is cached for 100 times as long as the execution time of the test itself. When a new test execution is requested, the cached value is returned and the test delay is avoided.

After a period of time the cache will expire and the results are discarded. This forces the execution of the test on the next request. The command 'tst forget' will force all tests to execute. This technique helps the runtime of the typical case but still has some exposure in the extreme.

Time to Fix a Failure Each time a test is executed, we compare the output of the test with what we were expecting. The differences between the actual and expected results are shown - they are calculated by using a 'diff' script. If the actual results match expected results, then diff generates no output and it is a passing test.

One way to fix a failing test is to approve the actual results returned from executing the test. This makes the current output the right answer for the next time. Here are a couple of examples:

```
tst like list-files

tst like code-length
```

If the test needs to be edited, the following commands might be used. This automatically runs the editor on the correct file.

```
tst edit list-files

tst edit code-length
```

It should be very easy to make small changes to any test in a few seconds. A lot of functionality can be added to these commands to solve other problems, but this is a perfectly useful starting point for your testing activities.

Frequency of Execution Given how easy it is to execute all of the tests and act on the results, you should be able to execute all of the tests at least every five minutes. When I am working in heavy incremental development the tests are executing every minute.

This testing style is also ideal for executing as a part of a continuous integration tool. Every code commit can kick off an integration text. Why not refuse to accept the code until the tests pass clean? This will prevent developer mistakes related to committing bad code.

Make your tests painless to run. If they are fast they will be run without requiring conscious thought or deliberate intention. Our goal with this strategy is to create a habit that becomes so ingrained that any alternative is unthinkable.

Test Cases

How can we create robust individual test cases that are easy to maintain over time? Our testing code will ultimately be composed of hundreds of individual test cases. Each test case should be incredibly simple and limited to a couple of lines of code.

Now we will turn our attention to some practical advice on how to build test cases for our system. We will examine how to structure a test so that we can easily filter the output. Then we will explore a couple of interesting applications for using our Diff Testing system. Common testing scenarios are needed to support testing with live data and testing web servers - these both require some special attention.

Initial Construction There are two levels of structure in the diff test implementation. The highest level of design deals with Test Suites. These provide for the testing of an entire subsystem or a specific type of data. The test suite acts as a container for test cases. The suite can be executed by itself to perform all tests related to a specific type of data. For example, 'brain test' would execute all of the tests available for the brain data type.

The second level is where we find the test cases. Each test case is a function that generates some output and can be approved independently. The suite simply enumerates the available test cases and provides a context for execution.

Here's how to construct test cases that provide you with the maximum leverage opportunity. Keep your test cases simple. In most cases you already have code in your product that does the thing you wish to test. Simply invoke the function with a single call and print the returned result. For example, 'print(list_customers())' is a nicely designed test case - no extra fluff here.

Use this as a launch point for creating new test cases. Simple is always the best first step. The assertions that are common to other forms of testing are provided by the output checker. A test case is guaranteed to produce the expected result or it will be reported. This means that you can skip the step of identifying the correct output by simply approving the results that you are happy with and fix the others.

Filtering the Output Some output is not important to the success of a test. This output can be filtered so that it is ignored before the results are checked. This will remove certain kinds of variations that aren't of interest. Some actions will produce unexpected output every time they are run. There are several design patterns that assist you in producing tests quickly that strike the right balance between rigor and maintenance burden.

The text that is captured from the output of the test can be run through a filter that removes the text that varies from run to run. There are a number of different filters that help you handle specific situations. Each of these filters removes incidental changes from the output or transforms the output in a way that you can use as expected results.

- **no_output** - discard all output so that this test can never fail
- **text_filter** - remove text string of a pattern from the output

- **line_filter** - remove lines that contain the pattern text from output
- **line_limit** - set a range of lines that the output must have
- **path_filter** - eliminate file paths for the test directory
- **time_filter** - remove time and date info
- **replace_filter** - convert text pattern to string
- **match_filter** - save only the lines that match the pattern

Each of these functions lets you post-process your test output so that you can quickly construct the test that you really want. Here are some examples of how this might be used in real life:

```
def customer_test():
    line_limit(list_customers(), 10, 15)

def spin_up_servers_test():
    no_output(spin_up_servers())

def list_files_test():
    time_filter(system('ls -l'))

def list_users_test():
    line_limit(match_filter(list_users(), 'seaman'), 2, 4)
```

This approach is clean, simple, and produces code that is quite readable. It reveals the true essence of what is being tested. It allows you to ignore the problematic output while preserving the rest. Imagine the expressive power that is now at your disposal!

Testing with Data The most difficult problems encountered during testing involve how to manipulate saved state. I believe that this explains the rising interest in functional programming. If a block of logic is stateless then it is extremely easy to test. Every time you use a stateless function you supply all of the data as arguments.

But in the real world most of our systems are extremely "stateful" so we need to develop patterns for testing the state in our systems. One common pattern is the setup/teardown pattern. Each test run begins with a setup function that creates all of the required state. In some cases, such as the Django unit test framework, entire databases are created from scratch. The initial state is then transformed by the execution of all of the test code. Because the system starts with a known state the code can verify the validity of a new state. After all of the testing is completed the system can be restored to a good state by the clean up function.

To effectively use this pattern you must be very clear about the level of state change that occurs during set up and clean up. You also must figure out if new data is inserted into the system incrementally or if the entire state is reloaded from storage. For example, if the database is reloaded from a file the operation will destroy any state modifications from the previous data.

Tests can also create side effects in the system state. Tests may be coded to algorithmically insert and delete data. This requires that more careful thought be given to the side effects of the tests but it also opens the possibility of using the existing state as a starting point.

Choose the techniques that work best for the unique situations that you are facing. Build standard solutions to common problems and get everyone on your team to use the best practices. Your overall testing should include everything from heavy destructive testing to subtle and incremental checks.

Test fixtures allow the system to advance instantly to a known state. Fixtures are often encoded as JSON data and loaded into the system. Instead of replaying the transactions in a story, you can jump to the end result. This lets you set the starting state on the system, add some new transactions, and then check the ending state.

For example, "save game MyGame" and "load game MyGame" could allow us to save the entire game and recall it from a file. Now entire scenarios can be tested with live data on the live system. This pattern can be encapsulated so that it becomes easy to load state, run tests, confirm state, and restore state. Engineers can then focus on creating the scenarios that need tested.

In a real system you need a robust server strategy to safeguard against corruption of production data. Several tiers of servers is essential:

- Production - Enable passive monitoring and notifications of errors but never allow live data testing
- Staging - Identical to production except for the fundamental connections to customers
- Test - All servers used to test the software
- Dev - Servers used to by engineers to develop and test the software

Make sure that everyone in your organization understands and follows the guidelines for each server environment. I recommend that you put automation in place to enforce the correct behavior rather than relying on the discipline of the individuals.

Testing Business Logic There is a huge variety of tests that you'll need to construct throughout the course of a project. The design goal of diff testing is to give you a common set of tools that you can configure to meet the specific needs of any particular situation.

There are some common situations that you will likely encounter as you begin to dive into this new style of testing. Develop a catalog of design patterns that represent the solutions that you face in your unique business situation.

Identify the key interfaces throughout your system. Learn the language of each API. Implement a single test case for each operation in the API. Create a test suite for the API and include all of the test cases related to the API. This makes for a simple but powerful way to organize all of your testing.

Identify all of the key data that is either imported on exported from your system. If you import and export an employees data type from your system there should be a test that looks like:

```
def import_export_test():
    print(import_employees('bogus_employees.txt'))
    print(export_employees())
```

This will let you keep an eye on the import/export operation without conscious effort on your part. If Wally accidentally breaks the test case three months from now you will know.

Web Page Testing and Browser Automation Web development is an important part of most software these days. You need an easy way to confirm that your web servers are working correctly and that the application logic is also operating correctly. This testing can be done either by accessing the web services directly or by remote control of a browser.

Curl is a simple tool that will allow you to make http requests directly to a web server. This bypasses all of the interactions that may be performed within the browser. There are options for things like logging into the web site and passing input to pages but these constructs are fairly primitive and I recommend against building a lot of testing logic using Curl. However, as a page fetch engine it will let you quickly create simple tests.

Selenium is another useful tool that acts as a remote controller for a web browser. It works on top of all of the popular browsers. You can use it to fetch certain pages from the web server and find certain elements in the DOM. You can even script activities like logging in and entering text in a search box.

Selenium is the best tool for testing web servers through a local browser. Because Selenium runs on the client, no special software needs to be installed on the server. This means that tests can be run from anywhere. A simple test can monitor and ensure that the production server is up and running.

Once the infrastructure is in place it is easy to keep expanding this testing. Scripts can contain logic to detect a wide range of errors that might occur in your application. Your only limitation is your imagination and the time you are willing to invest in contriving test scenarios.

Pages can be fetched from the server so you can easily compare either the text or the HTML of the page. An automated tool like this can easily fetch 100 pages and find the three pages that have errors on them. You may choose to fix the code or just approve the answer that you just got as the new right answer.

Selenium can also be used for animating the Java Script controls for the front end of an application. Many apps are becoming increasingly complex on the front end and Selenium provides tools for testing out the interactions of controls with heavy Java Script functionality. If you are going to do a lot of this type of work you might want to investigate other tools that specialize in this area.

User stories define the specific scenarios that users are trying to accomplish when they use your system. Consider using Selenium as the basis of a test app that will walk through each of the scenarios. The app can fetch a page, login, click on certain controls, view tables, and other page content.

Example Test Scenario:

- Go to home page
- Login as "Bill Gates"
- View product selections
- Click on order product
- Fill out shipping form
- Submit order
- View "Thank You" page

Overall, Selenium is a great tool to have in your arsenal. It is worth investing the time to learn how to make it useful in your context. If you do a lot of web development then Selenium can help you automate most of your manual testing.

Test Suites

Most testing advice focuses on tactics and there is quite a lot written on the mechanics of testing. My goal is to focus on the strategic nature of testing and how it can help you reuse software more effectively. I recommend that you get familiar with tactical testing techniques as well. A great place to start is with "Test-Driven Development" by Kent Beck, this is a classic that every engineer should be familiar with.

Testing strategies are the most important tool that you have when working with legacy code, yet they are often overlooked. Engineers often spend too much time thinking about the code changes that are needed and not enough time thinking about how the changes will add risk to the system. Build test suites that implement the different strategic goals you have for testing.

Testing Strategy All test code should begin by capturing the exact behavior of the current system. Implement some code to record the functional surface area of the app. Generate a series of requests to your system and collect the various responses. Think about hundreds of request/response pairs.

A single test case is nothing more than a request and its corresponding response. How these test cases are executed will be a function of your system architecture. There are numerous ways that you could implement these interactions:

- Function call that returns a value
- Method call on an API
- HTTP request and response
- COM or Corba invocation
- Web service query
- Database query
- Key/value pairs
- Script invocation with an output

Decide on what makes sense for your system. Do the thing that is easiest and most natural for you. You will rest a lot easier if you know that you have a library of hundreds of test cases that you can run in a few seconds. Consider building unique test suites for each type of interaction with your system. This has the advantage of utilizing your system in many different ways.

For example, you may wish to have the following test suites:

- database - execute stacks of SQL commands
- web-requests - deal with HTTP requests and responses
- import-export - consume and produce text files
- key-values - verify interactions with Key/Value store

By having different suites you can make them very simple and yet flexible. Each system invocation can use the same helper functions to call the appropriate entry point and process the output data in a consistent way. This scales extremely well to large system sizes.

Build simple test cases, rather than complex ones. Don't build test cases with multiple clauses that are trying to check for multiple things at once. Instead, build many different test cases and set these up as tables. Have a tool that will execute everything at once so you can run all of your tests at least once every hour. Run extended tests at least daily to ensure that you didn't miss anything.

Think about the types of testing that are needed to provide you with confidence that everything is working as it should be. Budget your testing effort to focus on the most critical areas. Give special priority to:

- Areas that need a lot of new functionality
- Performance bottlenecks
- Parts with a lot of issues historically
- Sources of current defects and other issues
- Costly parts requiring extra development time
- Areas of disagreements within the team
- Modules with many changes or multiple authors

Consider creating a test plan that has a table of all modules and your key attributes for making decisions about the priority of each. Create a scoring scheme that matches the business and development objectives of your organization. Score every module and use the results to set the development priorities. Select tools and techniques for each module that will be used to encode and execute the test cases. Now determine how many test cases you need to feel comfortable with that module.

Testing becomes strategic when it drives the rest of the development. Strategic testing is truly automated when it executes stacks of test cases without intervention. A table of results tells you which tests fail so that all failures can be fixed in seconds, not hours. Great testing should reveal local hot spots in the code.

Solid testing makes the code healthier over time by allowing constant refactoring - and refactoring can't be done safely without tests. Testing provides a strong indirect benefit to the long-term health of the entire system by turning back the effects of bit rot.

Test Execution All of the tests should run on a regular basis. If tests are only run on special occasions you will have to wait a long time to find out that the system is broken. It would be far better to find out within minutes of the system breakdown - shorten this delay to reinforce the correct developer behavior.

Tie the test execution into the commit process - continuous integration tools, such as Tox and Travis, can make this easy for you. Build and test the code each time you accept a commit to confirm that the code works properly across the full range of execution contexts.

Additional rounds of testing should occur automatically to ensure that the entire system remains in good shape. I recommend setting up multiple test frequencies: hourly, daily, weekly, monthly. Hourly tests should be limited in their scope and execute extremely fast because this is the basis of the developer tests and will frequently have a human waiting on the test results.

The monthly test, on the other hand, may take many hours to run. No one will be waiting for the results, so it doesn't matter how long it takes to complete. It may provision servers, restore databases, analyze transactions, and transform key data. In the limit, these "tests" may not be tests at all but key elements of your operations task automation. Define the different levels of testing so that they compliment the desired operations design.

A test suite defines a collection of test cases that execute together. Clearly designate which suites need to run in what frequency. Then match the speed requirements of the test suite to how frequently they are run. This makes it convenient to run a full set of tests in a few seconds while still ensuring that the more thorough tests are being run.

Working with Other Test Frameworks Like everything else in software development, testing has become increasingly heterogeneous. Different testing frameworks are optimized for different goals. It is often beneficial to use multiple frameworks together but this requires a clear strategy for how to manage the test code.

The most widely used testing tools are built on the unit testing frameworks that descend from JUnit. Collectively, the unit testing frameworks are usually referred to as XUnit. Diff tests can work well with the tests that have already been developed using XUnit.

XUnit tests are built in layers: module, class, method, assertion. Each level of abstraction provides containment for lower level constructs. The framework will automatically find and execute all tests and then present the results in a convenient way.

XUnit requires that all logic that you want to check be coded into the assertions of each unit test. In other words, you must decide what to check for as a result of every test and create an assertion that validates it. In contrast, Diff testing will execute any test and present you with the actual results. This allows you to approve the results after the test is written rather than before. More importantly test maintenance is done by verifying that the output is acceptable, not by figuring out what the output should have been. I find that this is a significant benefit of diff testing over XUnit Testing.

It is many times faster to fix a diff test than an XUnit test. The burden of maintaining an array of a thousand unit tests often gets so big that the entire idea of unit testing is abandoned halfway through the project. This leaves the project incredibly vulnerable - it would be better to have never attempted unit testing.

Diff testing is a welcome solution to any team that has struggled to successfully implement unit testing. It is easy to create new tests and quick to repair them. The typical programming session begins with lighting up the tests to see what changed. Ten tests fail and two minutes later nine of these are fixed and the tenth issue is something that needs attention. This case demonstrates the benefit of this testing style.

Traditional techniques wouldn't even find these unexpected changes because they aren't really errors at all. They are unexpected results that you should think about. Over the course of a year this system could save hundreds of hours of effort by giving you early warning of dangers.

The easiest way to utilize existing tests together with diff testing is to wrap the top-level test runner as a single diff test. I use a test runner called Nose to automatically run all my Python unit tests. Here is a diff test that will run all of the unit tests.

```
def nose_tester():
```

```
system('nosetests -v')
```

The output is automatically captured from Nose. It is compared to the last time the function was run. The "-v" option lists all of the individual tests. The names of all the tests run are confirmed. If anyone has disabled or added any tests the nose_tester will fail. With one line of code you can test the tester.

In practice, most tests invoke an external shell script rather than import code directly. This has the benefit of isolating the test invocation from the function under test. You will need to make a decision about whether you have a long running program with a test API or whether you are running different processes with each test function. Pick the architecture that works for your system but design it early on.

The most beneficial kind of testing is looking at the system context. This is often overlooked in favor of business logic. The implications of running out of disk space, and improper configuration of variables and paths, can be profound. Consider the following tests and ask yourself when they might have saved you.

```
def context_test():
    bigger_than (shell ('df'), 100000)
    line_limit (shell ('find'), 0, 20000)
    system ('du  thisdir')
    system ('echo $X')
    system ('ls')
    system ('pwd')
```

No matter how you do testing, it is a good idea to measure the test coverage. There are some excellent tools available for this. Search 'Test Coverage' with your language on Google. Push for high test coverage when you run your automated tests but accept the reality that you will not get to 100%. Use coverage information in addition to issues reported to determine where the hot spots of low quality are in your code. Analyze and fix areas of weak testing as you go.

Legacy Code Many of the books that discuss testing focus on building new systems from scratch. These are techniques that can be applied from the start of a project that will put the project on stable footing. If you have the luxury of setting the proper process in place at the beginning you will save yourself a lot of work later.

In the real world, most projects don't have that luxury. You will often inherit a mess that someone else created and you have to bring it under control. Let's look at testing techniques that are particularly good for dealing with legacy code.

A legacy system is any project that is currently deployed. It comes with a fully completed product definition, design, and implementation. It also comes with a history of trade-offs that were often made by a different team than the one that has inherited it.

Although the system is complete in one sense, it is also quite incomplete. It has issues associated with every phase of development, including the product definition, design, and implementation. Common testing techniques often don't work well with legacy code since they assume that you are starting a fresh application.

The first goal of working with a legacy system is to draw a functional and quality baseline. Evaluate the issues of each part of the life cycle to understand how the system works currently. This will help us determine the work priorities.

Once we have a baseline and priorities we will need to bring the system under control. Build some scaffolding logic and instrumentation around the areas of the code that need work. This will protect us from making errors as we begin to extend and improve the system. Don't skip this step in the process! Jumping directly into modifying code without a safety net will cause your software schedule and cost to balloon.

Our goal is to leverage the testing effort between projects. We intend to bring the full weight of understanding to bear on the new testing. Use product functionality that remains unchanged to identify the areas where the testing can also be recycled.

Identifying key interfaces within the system and create test solutions to validate them. Use playback and capture techniques to thoroughly test these interfaces. Analyze the API language of each interface in order to create a complete test suite. These interface tests are a key point of leverage between projects. Most interfaces remain remarkably stable over the years. Build sophisticated methods for testing scenarios on your most critical interfaces.

Best Practice #7 - *Use diff testing to generate maximum test coverage.*

Problem Many testing techniques require that tests are built in parallel with the code. Unit testing frameworks are often used to allow the developers to build tests for each function that they write. This can be a time consuming and tedious task. The developer's commitment to unit testing often wavers as the project progresses. Late in the project (when it is most needed) unit testing is often abandoned altogether.

The problem is that the testing techniques take too much time and energy to use effectively. Legacy systems also present a unique problem. The investment to retrofit testing onto an existing code base is enormous. This leads commonly to very limited use of Test-Driven Development techniques in legacy systems.

Solution Create some methods that utilize the ideas from XUnit but add tools that make it easier to write and maintain tests. By making every test emit output we can use **diff** to verify the output. Test cases can be reduced to single lines of code that simply invoke some function in our system.

Instead of having to anticipate and code for the correct test answer, each test can simply be written to produce an answer. On the next execution the test is required to yield the same answer. With this method, tests start out very noisy with lots of false positives, but can be easily silenced by using filters. The maintenance burden is then drastically reduced compared to other types of testing available.

Next Steps

- Create some simple tests
- Make a list of 100 assertions that you would like to test
- Plan how you might do live data testing
- Learn about Selenium for testing your web server

PART 3 - Leverage in Operations

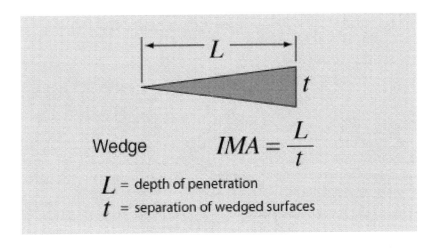

Figure 10:

Chapter 8 - Release Cycles

"To successfully implement continuous delivery, you need to change the culture of how an entire organization views software development efforts."

~ *Tommy Tynja*

Deciding when, what, and how to release the software is one of the most significant decisions that can be made on a project. Waiting too long will clearly hurt the business and lose customers. Releasing too early can also cause disaster for the project. There is clearly a sweet spot for when to release software.

What is the rationale for release? Do we just release the code because a date goes by and our shareholders are restless, or is there a better way to choose the introduction date? There are two major factors embedded in the software itself that drive this decision: functionality, and quality. The features must be at an acceptable level to meet the market demand or it makes no sense to release the product. Likewise, the quality must be above some level that is acceptable. When both of these goals have crossed a threshold it is time to release and any further delay will hurt the business.

Short release cycles will validate the next round of requirements using real customer feedback on new features - and you get fast turnaround on new requirements. Often the initial guess at requirements aren't what is actually needed in the product. Iterating on the implementation is the fastest way to get closure on the requirements. This must be balanced with the ability of customers to embrace the new features and give strong feedback. Rolling too often can frustrate your customers so it is important to minimize the number of releases that are in *simultaneous use*.

Controlling Scope

Unbounded scope on a project can spell disaster. The project scope is set by the breadth and depth of the problems to be solved. The feature set can be minimal or large and each feature can be simple or glorious. Larger scope means larger complexity in the resulting solution. If you solve lots of problems and create sophisticated solutions for each one, the project will grow in size. Complexity is non-linear in its impact. Doubling the scope will cause the required effort to double and sometimes the impact is far more. Scope, by nature, is unbounded because we can imagine many problems that might need to be solved.

Leading Cause of Late Projects   Scope creep is a significant struggle in software development. One more feature here and one better solution there adds up rapidly. Unless a budget is set and ruthlessly adhered to the project will inflate. No single factor has sunk more projects than trying to solve too many problems.

Each feature that is added must justify its worth. Every feature that doesn't add real value to the customer should be eliminated. This is half of the equation. The other half is that every feature planned is matched with a reasonable level of engineering. A project that is trying to build a million dollar system for a hundred thousand dollars will never leave the launch pad. Features must be carefully prioritized and be realistic relative to the investment.

Cycle of Increasing Scope   A long project cycle begs for a huge scope to cover the investment. If you believe that it takes ten engineers a year to write your web application you are planning to invest 20,000 hours in the project with a huge budget. There better be a giant payoff at the end of this project!

Because we can imagine a glorious system that will justify a huge effort we scope the project to be the greatest thing that we can imagine. A list of requirements is published that is twice what the team can actually produce. After several rounds of hostile negotiations between Marketing and R&D the project is reduced in scope to 150% of what the team can handle.

Act three of the play involves the developers actually building the huge feature list and discovering that the stated requirements were a small fraction of what was actually needed. This project will inevitably slip by more than 50% of its schedule. This results in both a functional and quality shortfall with a matching budget overrun. Unfortunately, this scenario happens every day throughout the software industry.

To meet this crazy schedule:

- development is rushed
- architecture is lax or abandoned
- the team loses track of exactly what is being built
- no one has the Big Picture of the project, even though everyone is committed to it.

Unreasonable Expectations The above scenario is far from inevitable. In fact, it is completely avoidable. The fundamental problem is there is no feedback to force an alignment of what is desired with what is possible. In this void, unreasonable expectations flourish. It is time to change the conversation entirely. The scope must be based on a calculation rather than a vague hope or wish.

The functionality and schedule must be posed as a question rather than assuming a fixed answer. Otherwise, projects end in a death march because the project release is over-constrained. A very different approach is required in order to produce better results.

At the beginning of a project, management typically believes that all of the product requirements are thoroughly understood. In reality, about 50% of requirements can be articulated up front and the rest must be discovered over the course of the project. This means that half of all problems that must be solved are not even known at the inception of the project. Often management starts by setting the end date, the core functionality of the system, and the staffing for the project. Given these constraints the developers struggle to produce the best possible plan to meet the requirements. The only alternative is to control scope throughout the entire project life cycle. It can't be assigned by the stakeholders at the beginning without jeopardizing the project's success.

Learning from History Look at your past projects. How close did you get to hitting the dates and feature set developed in the initial plan? Most projects disappoint their stakeholders. History will repeat itself unless your organization can learn and grow. This is where best practices in leverage are truly powerful. They can free you from this frustrating cycle.

Product Backlog and Burn Down Graph   Here's a few techniques that, if consistently applied, will put you on a path to never miss another deadline. Start by making a list of all the high level features that your project needs to implement. Work on it until you have around fifty different features listed. Now prioritize the features in the list by ranking every feature relative to the rest.

Now drop the ten lowest priority features. Does your project still make sense? If customers can get the primary value then it probably does. Now lose the next ten. Does your product work with only 30 features? If so, this is your minimum viable product. It must be viable and it must be minimal.

It is critical to manage the feature list as a single list. A common mistake is to negotiate features with independent stakeholders, but doing so is signing up to fight a losing battle. It is imperative that the stakeholders be managed as a single entity. Then the whole debate comes down to "what are you willing to pay for"?

As you implement the feature set, calculate your velocity. How fast are you moving through the list? If it takes you a month to knock off 2 of 30 items you can project that it will take you 15 months to finish (of course, the relative weight of each item needs to be normalized). Reconcile your desired rate with the actual rate of progress to ensure that you get to the finish line in time. If you are in trouble you must decrease scope or increase resources while there is still time to do so.

Managing Quality

You cannot, or should not, leverage work products that have poor quality. Cutting corners and taking shortcuts creates technical debt that will cost you much more later. Poor quality can have an exponential effect if reused. As you progress toward release make sure that you pay down all of the technical debt. A stream of low cost releases can only be achieved by a continuous investment in quality. Taking any path that sacrifices quality will ultimately lengthen the schedule of the current and future projects.

Service Level Agreements We typically set up Service Level Agreements to define the specific terms of software services we use outside our organization. They define the expectations for reliability and performance. They also specify what happens when things break down and who is responsible. Clarifying expectations in writing can increase the commitment to the shared goals.

A good agreement makes for good neighbors. If an SLA is a good idea between companies it may also be useful within a company. The internal agreement could define the terms and conditions related to how software is released into production. It could reference the specific numeric criteria for release and the testing protocols that must be in place. It could also outline the scenarios for promotions from staging to production and rollback if disaster strikes. This type of SLA can be the backbone of a release plan for operations. The day may come when this thinking will save you a lot of pain.

Quality Goals and Release Criteria When is software ready for release? A rigorous but lightweight process is necessary to green light a move of code into production. Our release criteria provides a checklist that can verify we have taken all of the steps to prepare the software and release infrastructure.

Customize this criteria to your organization's needs so it matches your business context. Here is a sample checklist to get you started:

- All defects resolved
- All tests pass
- User acceptance tests pass
- Stress tests pass
- Rollback tests pass
- 95% test coverage
- All features implemented
- No modules have complexity > 1000
- Developer review and sign off

- Operations review and sign off
- Review of all deferred defects
- All servers built

Avoiding Service Disruption When things go bad during an attempted release it can seriously undermine trust within the organization. The release criteria give us a first level of defense against bad releases but there are other things that should be in place as well.

There should be different types of server environments available with clear roles. This protects against accidental deployment or any other form of disruption within operations. Developers need to use servers that are completely outside of live operations. Test environment servers should be used to deploy all of the code and perform all of the automated tests. It may also be useful to do manual testing on the test servers. For the maximum amount of leverage, the test servers should be as similar to the production environment as possible.

Staging servers should be almost identical to the production servers. This provides the opportunity to do one final test before exposing our customers to the new software. Sometimes it a good idea to add in logic that detects what class of server the software is executing on in order to prevent certain operations from occurring. For example, imagine a system that sent customer records to another system. You might want to prevent this from actually happening unless you are on the production server. On the staging server it might just accumulate the records in a file. While this sort of test versus production encoding in the software is often inevitable, it must be minimized and very carefully questioned and scrutinized.

Reliability and Robustness When we plan to release a stream of products rather than a single release we leverage everything in our operations plan for each successive release. Gone are the days when it took months to release the software and years for the next product roll. With continuous delivery, it only takes months to get an initial release and follow it up within weeks.

This has huge implications for reliability concerns. An expensive release cycle limits how often software can roll. Therefore, reliability must be phenomenally good in order to release code continuously. High reliability allows minor changes to be injected at any time in order to fix critical issues. In this environment a round of bug fixes can be released within a week.

A good choice is to do a rolling update where the new software is introduced into some of the servers while the entire system remains running. Some of the servers are running the new software while others are running the old software. This is a very effective strategy to avoid any system shutdown and allows us to carefully avoid certain kinds of changes. Load balancers are used to route the traffic to the available machines and eventually all machines are running the new software.

If things begin to unravel the process can be reversed until someone can analyze what caused the failure. New software must be both backward and forward compatible. In a more traditional scenario, the system is taken off-line and all of the servers are updated, then the system is made available again. This has the advantage of being simpler at the expense of a disruption in service. Your operations agreement should cover these kinds of policies.

Most systems have state that varies over time. It may be useful to take snapshots of the running system periodically that could be used to reconstruct the system in the event of a disaster. This type of a design may be useful but keep in mind that these systems often have tricky concurrency issues that must be worked out.

Shorten Release Cycles

Over time software schedules have gotten progressively shorter. In the 1980s projects were measured in years and now the typical project can be completed in a few months. There are many reasons for this. Languages, frameworks, code generators, and other tools all combine to give the modern programmer an astounding productivity rate.

Higher productivity results in shorter projects or more frequent releases. To take advantage of the shorter cycles, projects must be very selective about the scope of the features that are delivered for each release. A short schedule doesn't allow for any work that isn't essential. It also requires that the release process itself be streamlined with automation.

Maximizing leverage is essential to meeting shorter release schedules. There isn't time in the schedule to create a massive amount of new code and get it functioning properly. Instead, the project must reuse the bulk of what has already been created and improve it to adapt to the new requirements. Shorter project cycles make it is easier to maintain focus on primary objectives which also improves the management of the project.

Shorten Your Release Cycle Traditional projects were measured in years while modern release cycles are in weeks. But how quickly can we really go? Start by looking at your current release cycle. How long does it take to go from starting to plan a feature set until it is being used by real customers? This is the length of your release cycle and includes the whole development life cycle including the time to promote the software into production.

Unless you have already been optimizing this, there is a long chain of serial tasks that culminates in happy customers. There will be many people in your organization that believe this is the fastest possible speed. In one sense they are correct, given your current process there is a minimum cycle time you can achieve. Our goal is to replace the current process with a better one.

It is entirely feasible to get to market in half the time, but you need to address all of the inefficiencies that are holding you back. To illustrate, assume that it takes you a year to create a product. This probably means that you have complex development, project management, and quality assurance processes in place that require several months of effort to complete a simple product. If this is your reality, your project must have a massive ROI to make it worthwhile.

Instead of a one year timeline, cut your schedule to six months. This will force your team to grapple with the biggest inefficiencies causing delays. It is very practical to do this size of a change in a single product cycle but you must rethink your ideas about life cycle and leverage. You will need to figure out how to leverage more and write less software. This will certainly lead to smaller scope and a faster release.

Market Demands Quick Turns The first dramatic change you will see after you implement these changes is in time to market. Customers will get new functionality in half the time. This doesn't have to be at the expense of functionality. The smaller scope and increased leverage will ensure that all core features are delivered. Usually, what is pruned off is the extra features that no one cares about anyway. If you follow this path the customers will receive higher value, more quickly, from the features. The customer experience will feel like you are giving them more, not less.

Reducing your time to market is a huge benefit for your stakeholders. Shorter projects will consume less resources. Parkinson's Law is the adage that "work expands to fill the time available for its completion". (http://en.wikipedia.org/wiki/Parkinson%27s_law).

Reducing Time to Market Improves Leverage Leverage results by reusing knowledge gained from solving previous problems and applying that understanding to solve new problems. A shorter schedule forces more careful selection of the problems and how we create the solutions. The shorter schedule demands leverage, which in turn saves time in the project.

Have you ever added people to a project only to have them slow you down? The new people do useful work, they just aren't useful enough to offset the cost of adding people. Removing people, or time, from a project has the opposite effect - you have less resources so you must reduce scope. A smaller job will be done far better than the larger one because of better control over scope and focus.

Whatever your current release cycle is, you can probably benefit from cutting it in half. After you master that level, do it again. By the time you get to one or two weeks you have eliminated all of the inefficiencies in the release cycle itself. Then the choice to release on a particular schedule is an operational one. Operations may only have a need to roll their software every other month. This shouldn't be an R&D limitation but an operational choice.

Tools for Continuous Delivery Achieving short release cycles requires you to rethink the integration. In a big bang project a lot of development goes on independently and then occasionally all of the parts are integrated together. This isn't feasible if you are going for maximum leverage because it creates a lot of extra debug work to integrate large pieces together into a whole system. Many of the tests that work for isolated parts will fail at the system level.

Instead, work to keep your code integrated together at all times. Prevent divergence and build everything incrementally. Only build small pieces in isolation to prove out design ideas but then integrate them immediately into the overall product. Write a full battery of tests that use the entire system in its integrated form. Otherwise, you are not really testing the integration, which is where many of the problems will lie.

How do you guarantee that the code stays integrated? By building automation into your core process. Every commit to the version control system should trigger a full system build and test cycle. With automation, alarms go off if the tests fail. This requires some initial setup but will save you countless hours trying to track down how the bad changes got into the code.

Make sure that you test in all of the contexts that your product will function in. I do most of my work in the Python world which has many different execution environments that may be important. For example, if I'm building a package that customers will use it should be tested for Python 2.7, Python 3.4, Pypy, Anaconda, and other environments. A tool called 'Tox' can create a series of virtual environments that will build and test in each of these environments.

Other tools can help you automate your integration process. Travis or Drone can invoke the build and test cycle with each commit. Travis monitors the Tox output and escalates any failures that are produced from the process. I heartily recommend that you set up something similar for your specific environment. Avoid the heartache of having to troubleshoot bad code by preventing it from happening in the first place.

Continuous Delivery

The only way to approach continuous delivery is by investing in automating every aspect of the workflow. Traditional development practices relied heavily on engineers to perform steps in a particular order to release a new version of the software. This is an inefficiency that software organizations can no longer afford.

Count the error-prone steps required by humans to get your software deployed. Exceptions require special activities to investigate and correct them, however, the happy path should be very streamlined. You know that the process is fully optimized when a chain of events can be triggered by a single command.

Automation End-to-End In order to achieve this kind of efficiency we must automate every task that doesn't require human judgment. Create a chain of events that can be scripted up to the point that the responsible person can approve further action.

Let's look at a typical workflow for automatic deployment:

Continuous integration

- A build and test occurs automatically before any commit is accepted.
- All versions of the product are built.
- Development code is merged into the correct branch and repository.

Test servers

- Test servers are provisioned to be similar to the production servers.
- New software is deployed and tested on the test servers.
- Test results are reported to the deployment engineer for validation.
- Any mysteries are investigated and repaired.
- Code is promoted to the staging branch and repository.

Staging servers

- Staging servers are provisioned to be almost identical to the production servers.
- New software is deployed and tested on the staging servers.
- Test results are reported to the deployment engineers for validation.

- Any mysteries are investigated and repaired.
- Code is labeled and promoted to the production branch and repository.

Production servers

- New software is deployed and tested on the production servers.
- Test results are reported to the deployment engineers for validation.
- Any mysteries are investigated and repaired.
- Code can be reverted if any problems occur.

Monitor After Release

- Monitoring is done to ensure a smooth introduction.
- The first 24 hours are critical for new deployments.
- Error recovery by roll-back should be practiced so that it is a viable option.

Each of these stages requires some human evaluation but the bulk of the work should be fully automated. Over time the automation may involve parts of the evaluation that were previously performed by people.

Best Practice #8 - *Build for continuous delivery of software and use end-to-end automation.*

Problem Long release cycles encourage unbounded scope. Because the project is such a large investment, the stakeholders continue to add speculative requirements to justify the cost. This is difficult to resist and results in a lot of unnecessary features that customers may not even want. The longer the project runs, the larger the desired scope becomes. This results in a destructive cycle of feedback that destroys many products and demoralizes teams.

Solution The fastest way to prevent this from happening is to intentionally shorten the project cycle. This drives careful prioritization of the essential functionality. The short deadline also requires mastering the disciplines that are good engineering practices. Once a company learns how to successfully implement continuous delivery they never go back to a long product cycle.

Short cycles also require fully-automated testing for the integrated system. Tools can be hooked into the versioning system to automatically build and test in all of the desired environments. Each commit can trigger a full round of validation to qualify the recent changes. A failure will trigger a rejection of the changes and forces the engineer that submitted the code to fix the issues. Using these practices will make it possible to release code quickly at any time.

Next Steps

- Measure your last three projects. How many weeks between releases?
- Plan what you would change to cut the release cycle in half.
- Review any recent slips in expected release dates.
- Assess the effort required (in days) for one production roll.
- Asses how well-tested is your rolling upgrade and your roll-back plan?

Chapter 9 - Services Architecture

Cloud computing is creating irreversible changes in how computers are used around the world.

Services that Scale

Cloud services are the backbone of software development. Whether you are developing software for small business or large enterprises it will almost certainly result in applications that run in a data center. These apps may be hosted in either and private or public cloud but the fundamental issues are very similar.

Corporate policies for data storage often drive critical architectural decisions. Cloud services must be built on sound security practices that protect the data integrity and access within the system. Many environments have very restrictive policies which make it so that interactions with cloud systems are closely controlled.

Installation and setup are vital to get the system properly deployed initially and managed on an ongoing basis. Because of the unique issues with application deployment and entire chapter has been devoted to that topic. In this chapter we will focus on the underlying architectural decisions that will allow your system to scale over time.

Building web services is perhaps the most common software task today. This produces some unique opportunities for reuse. We can create parts of a solution that are easily isolated from one another. Each of these services can solve a simple problem and may end up having a different life cycle than the rest of the services that we use. This promotes reuse by letting us repurpose existing solutions to solve new problems.

The rise of microservices has given us new design ideas. We can now think about building and deploying systems that are much more focused on solving a specific part of the overall product. We can then build more complex systems by integrating them together to provide robust solutions for our customers.

Each of the services are built to address a specific concern. By limiting the functionality of each system we can build systems that are far easier to construct and operate. Control and data flow must be designed to meet the business needs. The services paradigm makes scaling far easier than it has ever been.

As the number of users push the boundaries of the existing system new designs are needed to address the increased demand. The well-designed architecture built around services allows portions of the system to be replaced to accommodate the new demands without disrupting the entire system. Therefore, our architecture directly sets the leverage potential of our web services.

Scaling your Architecture Services are intended to work well for some number of users. As demand increases minor changes can be made to provide the required performance up to a certain limit. Eventually the demands of the new users can't be met by the current solution. New bottlenecks emerge as the system grows. A system that works well for 100 concurrent users will not work for 1000, and a solution for 10,000 users will not scale to 100,000.

Usage can be measured in requests per second that the system is intended to respond to. Every order of magnitude in usage requires that some fundamental elements of the design be revisited. Each of these orders in growth represents a technical horizon, and typically has a corresponding business horizon for the company as well. What is your next horizon and how quickly do you need to meet it? Some business are under tremendous pressure to scale while others are not.

This chapter lays out some principles and practices that will help you scale your systems. If scaling is not a major issue for you then you can skip forward to learn about other operational issues.

To build services that scale you need to find and eliminate the bottlenecks that are preventing your system from performing at the required level. Once a key choke point has been identified a new design must remove the bottleneck while not disturbing anything else.

Design for Leverage Reuse is an architectural concern. A poor design is very difficult to change to meet new requirements. Modifying one part of the design forces an understanding of the entire system and may break the software in unexpected ways. This is a design problem. In order to scale your services you must start with a realistic assessment of if your system can be scaled. Is the design sound enough to let you refactor portions safely? If not you are better off building a new system that leverages the understanding while rewriting the current system.

The preferred option is to scale your existing system rather than replace it. Scaling your current system means that you are working with legacy code. Everything discussed in the chapter about Design Leverage applies here. Architectural and development requirements include:

- Complete battery of end-to-end tests with high test coverage
- Core skill at refactoring
- Version control of all app and configurations
- Skill at instrumenting and performance measurement
- Effective development process to deploy code safely
- Robust server strategy: dev, test, staging, production

If you are lacking any of these key ingredients then fill the gap before moving on. Ignoring these issues can result in making serious mistakes that are difficult to recover from. When you have what you need it is time to begin the work of scaling to the next horizon.

The actual work of scaling starts with finding the next bottleneck that is holding you back. Is your application compute bound or I/O bound, which servers are backing up the requests, which queues are getting long, where is your maximum latency? Use analytic tools to watch all of the servers and how they perform. Find the next hot spot. This is your target for redesign.

Now, propose several different ways to solve the problem and select the simplest first. Prototype the proposed solution and see if it scales to the demands of the next generation. If it does then simplify it further. Build bridges and adapters to let you use all of the other parts of your system without modification. Once you are sure that this solution works then you can remove the extra logic for the adapters to get to the final solution.

Separation of Concerns Typical software products have a large number of different concerns to address:

- User authentication
- Authorization
- Web page routing
- Business application logic
- View rendering and template expansion
- CSS styling
- Dynamic behavior within views
- Object relational mapping
- Data persistence
- Database queries and filters

Each of these functional concerns should be implemented in a single place (as much as possible). If functionality is spread out across the system then changes in one area will cause problems in areas that are seemingly not connected. Can you point to a single place where each of these concerns is addressed?

There are some cross-cutting concerns that can not be entirely isolated, but they should be strongly encapsulated and consistently used throughout the system. This includes things authentication, logging, metrics, and caching. Although these concerns touch your system in many places they should touch it very lightly. A simple interaction should regulate the degree of coupling that the cross-cutting concerns have with the application logic. Otherwise you are asking for trouble when you attempt to leverage your design.

Fracture Planes

Every system has some natural boundaries within it. A fracture plane is a natural place to divide the design and impose boundaries for the design abstractions. A good architecture has reasonable divisions built into the design. On each side of the boundary completely different concerns are implemented and these are isolated from interacting with each other.

Find natural points of integration and build your architecture around them. Imagine independent life spans for each of the components touching the interfaces. Could you reasonably replace these parts independently without having to redesign the other part? If so then you have the right abstraction boundary. Evaluate cohesion and coupling of your subsystems and components. We are after loose coupling between the parts and strong cohesion within a given part.

Some boundaries within the system are due to dependencies in the technology stack. For example if one part of the code is dependent on the .Net framework and another it dependent on a Java library, then this may be a fracture plane in the design. Or if one part is built in Ruby and another in Python it may be a good place to split the design.

Architectural Layers All systems are built in layers that are characterized by different types of concerns. The software will eventually be deployed on a server somewhere. The cloud hosting services provided by Amazon Web Services or Google Compute Engine can be viewed as a service layer in our architecture.

The software will execute in some platform environment that is built on top of an operating system like Linux, Windows, or Mac. This in effect is another layer of our system. On top of the operating systems we build application logic that relies on particular programming languages. Each language provides us with access to libraries that must be set up properly in order to deploy our software.

Typical system layers:

- Hosting services
- Virtualization technology
- Operating System platform
- Programming language runtime
- Tools, libraries and frameworks
- Configuration and setup services
- Database engine
- Databases
- General tools for your company
- General app services for your company

- Application business logic that you wrote

It is important to understand the layers in your system. Every dependency may illustrate layering of some kind. The system dependencies will have a profound impact on the direction of the leverage that is possible. Your application logic is built on a stack of dependencies. Many of these dependencies are deeply coupled into your app logic. It is not possible to change a core dependency without rewriting all of the code built on top.

Replacing the lower level layers of a design typically requires a complete rewrite. If this must be done then make sure that you are prepared for the cost of building a new system. Scaling an application should not typically involve such radical loss of leverage.

Front-end/Back-end For the last twenty years most applications have been patterned after a three- tier architecture model:

- *Presentation layer* - implements the user interface
- *Application layer* - houses all the business logic for the app
- *Data layer* - creates persistence for all of the data within the app

Each of these layers has unique concerns. Select technologies that give you the best possible leverage within each layer. Wait for a good time to switch the key technology that you use so that you do not loose leverage in the transition. When you are building components from scratch is a good time to replace obsolete technologies with better ones.

Develop your own rationale for which tools you are moving towards and look for opportunities to move into them. Here is a sample of technologies that we use daily in my company.

- *Presentation*: Angular JS, Bootstrap
- *Application*: Python, Django, Django REST Framework
- *Data*: Postgres, Mongo DB, Redis

Benefits of Microservices Web applications used to be single apps that ran on a single server. Now there are typically many apps that run on managed clusters of servers. Each of the services is hosted on one or more servers throughout the cluster. Service- oriented architecture is the new normal expectation. A series of web services work together to accomplish the overall task for the business.

Recently, a new architectural model has emerged with the name Microservices. The design model favors fine-grained services each of which do only one task. The services are typically connected using REST APIs to talk to one another to coordinate the overall product goals.

Microservices encourage more leverage by making the coupling between the service extremely loose. The REST API isolates details about each type of business logic to a single service. For example, a system that tracks inventory would contain a single service that would know how to represent the list of parts. Other services would know how to request changes but wouldn't understand the way that parts were represented.

The application is built around simple services that are aggregated together. Any service can be replaced with a similar one. This makes it easy to replaces pieces of the application constantly without risking the entire app. Leverage abounds as every piece takes on its own life cycle. Scaling happens by locating the current bottleneck and fixing the service that is impeding the traffic flow. Microservices provide a path to much smoother scaling.

Data Services - Architecture and Partitioning Almost every system that you are likely to build revolves around handling data. Since this is such a significant part of modern software design you should devote a significant portion of design time to your data handling architecture.

If your problem is primarily about the relationships between data types then a relational database may be a good choice. On the other hand, if your primary concern is scaling then NoSQL solutions may be a better fit for you.

Another concern about designing persistence for your data is the boundary between the code and the database. Every piece of data may need to have two representations: one as objects in memory, and another as rows in a database table. You may want to think about how to use an

ORM (Object Relational Mapping) to convert seamlessly between these two data encodings.

Think about building reusable data solutions rather than rebuilding them from scratch each time. Plan how the structure of your data schema will need to change between your different systems. Because data is so central it can either be an enabler for leverage or its greatest threat.

REST Pattern The design pattern for RESTful web services lets you build strong key abstraction based upon a set of commonly understood trade-offs. REST is short for Representational State Transfer. It is the most widely accepted design idiom for encapsulating a type of data behind a web service.

REST systems are built around data types. For instance, the CRUD operations are typically implemented by using different HTTP methods:

- Create - HTTP put
- Read - HTTP get
- Update - HTTP post
- Delete - HTTP delete

The REST pattern dictates how to use the commands to access and manipulate the data that is held within the REST service. For more information go to the wikipedia article at https://en.wikipedia.org/wiki/Representational_state_transfer or consult the definitive source book from O'Reilly, "RESTful Web Services; Web services for the real world", by Leonard Richardson, Sam Ruby.

To create REST services that maximize leverage you should start by building around the most significant nouns within your system. Some data types are core to your domain and these require the best encapsulation. Then define the CRUD operations for each of your core data types. Add other operations that are required to implement your business needs.

REST is well understood. Your goal should be to leverage off of that understanding to avoid building your own solution from scratch. I recommend that you consider building a solution based on one of the great web service frameworks. My favorite tool is the Django REST

Framework. A skilled developer can build a prototype of a web service in a day. It has built in solutions for user authentication, advanced security, and automatic generations of all the interface code.

All you need to do is define the specific objects that you need to represent and the framework will generate all of the code for you. The code includes the API logic with authentication and even a testing GUI to interact with your object types.

As you move into the world of automatic code generation, it is critical that you build a testing infrastructure in parallel with the product. It does you no good to automatically generate 10,000 lines of untested code. Invest in an infrastructure that lets you create test code that contains your business logic. More on this is the next chapter.

Evolutionary Architecture

Building architectures that scale is straightforward when following a few simple rules and applying them rigorously to each new problem.

- Keep your design simple
- Keep your code clean
- Keep your code integrated
- Keep your code tested

Problems in your architecture emerge when you relax one of these requirements. Over time, a small issue can cause bigger problems if it is not addressed. Regular team reviews of the code and design will keep things clean and tidy and reviews will also spread the information around within your team.

Strategies That Work At this point you probably have an easy way to measure the overall complexity of your system. This will become an important tool in your planning. Don't continue to make random changes without being able to see the effect that those changes are having on system complexity. Remember, system complexity is a major driver of cost. Each day strive to lower the complexity.

Your complexity measure should also give you a number for the complexity of each part of your system. I recommend that you track the complexity of your top subsystems and measure the quality of each subsystem. This creates an accurate picture of the top hot spots in the application at all times. A hot spot will have the most defects outstanding and the highest complexity. In my experience, these two things always go together. Don't give bugs a chance to breed.

The Services Paradigm Find the natural boundaries that separate groups of functions. Within the subsystem the cohesion should be high. In other words, all the parts of the subsystem probably talks to each other, while the subsystems are only loosely connected to one another. The subsystems should be loosely coupled, but internally well-connected.

Plan out how you can strengthen the architectural design of these interfaces. For example, two subsystems that are connected through a scripting language are bound to be much looser than if functions are called directly.

Clean and simple interactions make for a good design. A limited vocabulary on an interface lowers complexity. Clarify exactly who calls who and what they say. Reduce the amount of knowledge that different subsystems have of each other. Build great components into web services by finding those blocks of functionality that belong together and provide a REST API for the service. These types of services are very easy to scale over time because they are nicely isolated from other services.

Reusable Services and Integration Review each group of related data types. Data types tend to cluster around the functionality of your problem domain. Identify these clusters and build web services around them.

Build services with canonical interfaces. Use the common verbs to help you identify common operations that might be expected. For example: Create, Read, Update, Delete can often be applied to help define a nice canonical interface for a set of nouns. Another common use model is Open, Read, Write, Close, defined for use with files but the applications can go far beyond that.

Produce the types of interfaces that other developers would expect to see. Follow the principle of least astonishment. If they expect something, consider providing it.

Maintain the integration of the entire system. Don't let pieces remain in isolation for very long. They will be difficult to integrate with the rest of the system unless they are held in a state of continuous integration. While integration is important, so is fragmentation. Each piece must operate successfully on its own. This prevents unwarranted dependencies from growing like scar tissue between the components. Both integration and fragmentation of the entire system is required for maintaining the overall health.

Standardize on service interaction models. Use common, high-level design patterns to orchestrate the traffic flow between services. Identify reusable techniques to manage transaction processing, queue management, and caching. All of the high-level designs make it easier to scale your system and there are off-the-shelf tools that will assist you in building out and managing the system.

Managing Performance

Study system performance on the production servers in your live operation. There might be some surprises waiting for you. There is no substitute for analyzing the actual system performance. Study each of the systems to understand the overall traffic flow rates. Measure typical and maximum latencies to understand where the bottlenecks reside.

Examine how data architecture affects your performance. Imagine different ways that your data might be partitioned across the system. Create some alternate models for your data architecture and try to create experiments that will validate your assumptions about what is really happening.

Implement caching wisely. Overall performance can be affected dramatically by how you use caching. Improperly configured caching can destroy your throughput while proper caching can have a huge benefit.

Track key performance metrics continuously. Use the experiments to understand performance as a way to find the strategic locations for ongoing monitoring. Then leave the best tap points in place to track the performance on a permanent basis. Convert from doing performance analysis to performance monitoring. Set thresholds for notification if certain measures go outside the normal bounds.

Capacity Planning   Let business drive your development. As traffic increases, figure out how to scale your operations to match the demand. Don't focus on problems that might be in the distant future but start soon enough that you can complete the work before impacting the perceived responsiveness of your system.

Respond to the current reality by honestly assessing how you are doing in the eyes of your customers. Create a balance so that you don't overbuild or under- build your product solution. Avoid solving potential problems until they are actual problems that need to be addressed.

Identify the key bottlenecks that cause performance concerns and eliminate them. These will typically be related to how the services cooperate with each other. The most common areas of concern are related to authentication, caching, data, locks, and traffic routing. It is rare that a single service is the bottleneck because it is too slow.

Data is the Key Concern   Most parts of an application scale easily to multiple servers. Data is the central player in most applications so how data is handled can have a huge impact on the overall performance. As your app becomes more complex you should hire dedicated data architects to build solutions that will meet the demand.

Data may be replicated across many database servers and many databases. The data may be partitioned by sharing it across many systems and geographies. These are tricky problems and you should have highly skilled people that have spent years thinking about the relevant issues working on them. But for smaller systems you should focus on simpler solutions that will still meet the performance requirements. The simpler the solution the more leverage you will get from it.

Choose an architecture that matches your business problem. Make sure that the data queries are implemented efficiently. Measure the time required to do queries to make sure there aren't any glaring issues in your database layouts. Most databases provide tools for validating the data table structures.

Use load balancers to scale the number of users that are served by a single server. If you are running clusters of database servers make sure that the transactional load is evenly distributed across the servers. An idle server in a busy cluster isn't contributing anything to the performance and may be caused by a problem in the data architecture.

Performance and Caching Caching works well if the data is static. If each unique query can produce a single answer then it is a candidate for caching. For example web pages, CSS files, JS files, and images should all be managed through a cache. This can have profound speed implications and allow you to scale many time over.

Beware of caching dynamic data - not all traffic is stateless. Certain requests can't be cached reliably and so the cache must be configured to ignore these items. Caching does introduce complexity into the system but if you need to scale this is the first place to look for answers.

Be prepared to invest serious engineering into analyzing and fixing performance problems. Understand both the reliability and performance implications of when and how to flush the caches for each different type of data being tracked.

Key Performance Metrics Track the average and maximum number of concurrent users on your system. Learn what days of the week and times of the day are most likely for peak traffic. Design your performance goals around the peak traffic points. Also track the peaks over time to get an indication of how fast your usage is expanding. This will help you set reasonable goals for your capacity planning.

Measure the average and maximum response times. Understand the correlation between the number of users and response times. This will tell you a lot about how much margin is built into the system or how close to the edge you are.

Watching the system load each service is another indicator of margin within your operation. Which server types are the busiest? Adding more capacity here will speed up your system while adding capacity to idle servers will have no impact. Utilize specific measurements to calibrate your mental model for how your system is behaving in the real world.

Best Practice #8 - *Build your system from loosely connected services for optimal flexibility and reuse.*

Problem It is very difficult to develop and maintain large applications that are built as a single lump. Applications should be built in layers that match the functional needs of the problems that are being solved. Applications should be built as a set of services that interact with each other.

Poorly designed services are too large or too interconnected to allow any significant leverage. If it is difficult to make changes to one service without breaking something else, the maintenance of the system will be exorbitant, leading to an early replacement of the system.

As more users are added the system must be scaled up to meet the new demands. This requires active modification of the software. A system that can't be maintained for a reasonable cost will end up being replaced.

Solution A well-defined system is built in layers and multiple services interact with each other to accomplish the customers' goals. Because the services are independent of each other they can each be optimized without breaking other parts of the system. This makes it easy to make changes required to meet the increasing performance demands.

As we scale the system, special attention is given to data architecture, caching, and load balancing between the servers. This clean architecture can evolve over many years to meet new challenges.

Next Steps

- Identify the specific distinct layers in your system.
- Identify the unique services used.
- Map which concerns are addressed by which services.
- Analyze your interfaces. Would REST be a benefit?
- Measure users, response time, server load.

Chapter 10 - Application Deployment

> If you aren't embarrassed by the first version of your product, you've launched too late.
>
> ~ Reid Hoffman

Versioning

Version control is critical to the entire deployment process. Without proper management of the recipes for deployment, operations relies upon the staff to faithfully remember every detail of how to keep the system running properly. Humans aren't accurate enough at remembering a long series of detailed actions. This makes computer automation critical for smooth operations.

Version control is absolutely essential to automation. Each time a set of tasks is executed it must produce the exact same result. This requires writing software for every task, encapsulating the entire history of actions. A version control system must be used to allow explicit references to that history. You should be able to retrace the exact steps of what happened last Tuesday at 10:00. Without this degree of rigor your operation will become extremely unpredictable.

There are several good version control systems in widespread use. For the purpose of this discussion I'll refer to all of these as Git. You should be using a tool that has the same attributes as Git (such as Subversion, or Mercurial).

If you need to manage binary content, git may be inadequate. You may want to look at tools that are specifically set up to track binary content. Archiva, Artifactory and Sonatype Nexus all integrate very well with Maven, Gradle, and other build tools to download versioned binary tools. Artifact repositories are a leverage enabler.

All automation should be built on top of Git. Control the server environments and configurations by using branches. Merge branches together to manage the project work-flow. Make sure that all of the deployment steps are built into fully automatic tools. This will ensure that all deployment is done exactly the same every time. This gives you a history that can be repeated to track down and debug failures.

Everything is Versioned There are many types of files that must be versioned. It is important to distinguish between the following items, all of which need to be tracked and versioned:

- Source. This includes configuration templates or metadata
- Generated artifacts (from source). This needs to be versioned and stored in an artifact repository like Artifactory.
- Leveraged artifacts and tools. These also should come from a versioned artifact repository like Artifactory, which conveniently acts as a mirror of remote repositories
- Configuration files for our artifacts and leveraged artifacts. These are generally text files that are often generated on the fly by configuration management and deployment tools.

Proper use of Git requires that you truly understand the source code in your project. There are many files that are combined together to create a product. Some of these files are built during a build process. By definition, these are not source code since they are built by you. Therefore, they should not be managed within Git. Use *.gitignore* to skip over these files.

There are also tools that are required to build your project. Any tool that will not be readily available (by version number) should be added to Git. Don't take a chance at not being able to rebuild your product due to a missing tool. A lot of the tools that we use help us set up and configure new machines. Any files that are necessary for Configuration Management (CM) must be a part of the source code for your project.

The specific versions of tools required to build your product should either be tracked within your Git repository or available from some official source that is guaranteed to be around two years from now, or as required by your company policies. Make a conscious decision about these things rather than assume that something will be there. Create reproducible results by tracking everything that you need to build the entire world from scratch.

Standard Branch Policy Using branches within your Git repository can clarify many details related to work-flow. Throughout the development there are different uses for branches.

- Experiment with new ideas that may be incorporated later.
- Stage the integration of different pieces of code that rely on each other.
- Isolate responsibilities (eg. production, staging, test).
- Use branches to stabilize changes during releases.

It is important to have a solid branching policy. This helps all of the engineers know how to work together in an effective manner. The first policy to establish is how to slow down development close to a release so that unintentional features or untested code is not accidentally released. A numbered or named branch should be created for each major and minor release.

This allows the master branch to continue moving rapidly with lots of commits while the release branches begin to be restricted. Only changes that are properly tested and reviewed are added to the release branch. After the code enters production the rate of change should slow even further. Using a branch allows some change to occur but this should be tightly controlled.

In addition to release branches, it is a good idea to use branches to establish responsibilities for the work-flow. For example, Staging, Test, and Dev branches can be used to mark code that is at different quality levels. As more testing is applied the code is promoted to the next level by merging the branches together. This lets you establish gateways or thresholds that must be passed in order to go to the next level.

Branches are also used to experiment with ideas that aren't quite ready to put into the mainstream development. These may be built around specific features or used as integration branches for mutually dependent components that rely on each other. It is best to integrate changes before they are merged into the master branch.

Requirements for release branches

- master is point of integration for new development

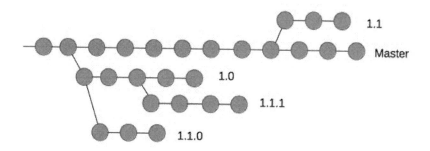

Figure 11: Git Branch Graph for Releases

- stabilization branches slow rate of change
- release branches subdivide for each minor release
- release branches are often merged back into master

Release Labeling Branches capture a series of commits. Labels mark a single commit as being significant. Every time the code changes in production it should have a label. This lets you know exactly which version of code was in production at any given time. Standardize the product labeling policy so that it remains stable over time. Create a new label for each production version.

Example labeling strategy - Release 1.2.7.100

```
1 - Major release
1.2 - Minor release
1.2.7 - Release number
1.2.7.100 - Build #100 for 1.2.7 release
```

Note that this would be tracked by two branches (1 and 1.2). If you would expect there to be multiple commits on 1.2.7 then you would add a third branch. The head of each branch should be set to the most recent label that is applied. For example, if labels "1.2.7.100" and "1.2.7.133" are used then the head of 1.2.7 branch would point to "1.2.7.133" label.

Release Testing A major part of the cost of releasing new software is related to the cost of maintaining adequate test server environments. The test configurations should be as close to the production environment as possible to provide a valid indication that the code will function as expected in production. Maintaining the testing environments can be a nightmare because the world is constantly changing. All of the tricks that we have been discussing apply to maintaining the test server setup.

As development proceeds, each commit is fully tested by using the automated testing on each iteration. This gives us some confidence that the code is working. Ultimately we need to deploy and test the code in a much more realistic environment before we are ready to trust it enough for a production release. This can be rather costly and involves configuring test servers that can do the detailed testing. When the testing succeeds on the various test servers the code is promoted to staging servers that are configured to be almost identical to the production environment.

The code is then labeled and promoted into the production environment. As the code begins to roll out through the production clusters, careful monitoring is done to ensure a smooth roll. If anything unexpected crops up then a roll-back is used to restore the previous state of the software.

An "After Action Review" should be done to study the results of the release. The team should get together and discuss what happened and plan any changes that should be made to the process before the next roll.

There is a minimum amount of effort required to fully test a release. There are computational and staffing resources that must be applied. Rebuilding environments or maintaining multiple environments is very expensive, even with extensive use of configuration management and deployment tools. This creates a minimum cost for the release cycle itself in addition to the cost of building new functionality into the system. This is the single biggest reason why organizations minimize releases and patches.

Measure this cost for your organization to determine the effort required for one release. Discover the sweet spot for your testing and release cycle. Then build your project release cycles to match your entire process.

Continuous Integration

Delayed integration is costly; stay integrated as much as possible throughout the development. The best way to do this is to make it impossible to submit code that is not tested. As features are created new tests must also be created to ensure that the features work properly and that they integrate with the other code dependencies.

Tools can enforce the desired integration practices. Each time a commit is attempted the code is automatically built, deployed, and tested. Only when all the tests pass will the code be committed to the Git repository. This prevents bad code from being committed. Of course, the effectiveness of this approach hinges on the robustness of the tests that you write. Wimpy tests may still allow bad code to be submitted.

Build tools, such as Archiva, Artifactory, Sonatype Nexus, Maven, and Gradle, all assist in continuous integration through the setup and control of building and testing code. Triggering these tools with each commit gives you a great degree of confidence in the code.

Use Tools to Enforce Integration There are several high-quality tools that make your job easier. They will trigger a build/deploy/test cycle by bolting into Git. They allow rules to customize the actions taken and produce notifications that keep you informed about the progress.

Here's a list of tools for further investigation:

- Travis
- Jenkins
- AnthillPro
- Atlassian Bamboo
- Microsoft Team Foundation Server
- Drone

These tools require some energy to set up but can produce many times the benefit on the first project cycle. Imagine the hours saved if you never had to track down bad code that got committed through carelessness. Each time code is committed to the master branch it initiates a testing cycle. Only code that passes all tests is allowed. All other code is left for the developers to properly debug.

Our projects may contain components that require different versions of the target code to be built and tested. For example, we might need to produce binaries to support multiple execution environments from the same source code.

- Different operating systems
- Different mobile devices
- Different product configurations
- Different server environments
- Different browsers

A tool such as Tox can be used to iterate over each environment and build, deploy, and test the code. In order for Tox to pass, code must run properly in all configurations and environments. The more elaborate environments require dedicated servers for managing all of this testing. Over time you can build a very robust test battery to cover everything that you truly care about.

Automatic Integration Integration is not a special step. If it is not viewed as a natural part of development then it will be postponed. This will cause problems later on as people begin to defer more and more of the integration work. Eventually integration failures will block progress. By investing in tools for Continuous Integration you will prevent this from happening.

Integration problems should be investigated as soon as they occur. If two pieces of code should be working together and they aren't, this is a high priority for debugging. This eliminates problems before they grow. New tests should be added to address any system weaknesses that are uncovered.

Place the burden on each developer to get everything working properly rather than some central staffing responsible for the build. If a developer makes an error that causes problems for others it should be their responsibility to fix it. Developer accountability is vital for the ongoing effectiveness of the entire team. Use integration branches to assemble pieces of code that might take several attempts to integrate. This protects everyone else's time. Some changes need extra coordination and integration branches may be the best answer.

A counter-intuitive approach to automatic and continuous integration is to make the development team, or individual, responsible for automating and monitoring the state of the integration tests. In this scenario you trust your teams to be entirely responsible for their product. They decide when it is "good". Teams themselves build their integration environments (making generous use of Docker or VMware workstation images, for example) on their development stations. If you are not testing massive scale, this is how your organization scales. For massive scale testing, you have a dedicated environment.

Off-line Emergency Drills Practice emergency response skills. Conducting "fire drills" is a way to ensure that everyone is prepared when the real thing happens. If you know that you must deploy an application from scratch you should practice this drill as part of your regular training regimen. Don't wait for the actual event to determine all the tasks necessary to bring up and configure clusters of servers and get them talking to each other.

You must have automation that goes from bare hardware requirements to running applications. This process should stop just short of deployment on the production servers. Practice setting up machines from scratch, data rollback, and restoration.

Build in test hooks to your servers to allow you to observe critical aspects of your running application. You may also benefit from special settings that are applied only on your test configurations. These can prevent accidentally doing actions that are only allowed on the production servers. Another approach is to make certain additional features available for debugging only on the test servers.

Be sure to provision dedicated servers for testing and staging. Don't try to cut corners here. Your production environment should be fully disconnected from your testing and staging environment.

Configuration Management

Configuration Management is about defining the server execution environment. It controls everything from the operating system version, application packages that must be installed, configuration settings for each component, and network configuration options for allowing the services to communicate with each other. Everything about your system is controlled by your CM.

Automate server setup tasks completely. You should be able to go all the way from your version control files and the hosting accounts to running applications by using automated tools. Many smaller organizations use scripts to provide this automation. Using leading edge tools can save a lot of work. These will generate solutions quickly for the short term and provide greater capabilities as your needs change.

You need repeatability in how your servers are configured. People make mistakes and it is dangerous to rely on the knowledge of staff to understand how your systems are configured. This must be fully automated and based on CM files to be repeatable. Configuring machines properly is probably the hardest problem in deployment and operations. The state of the art tools (Chef, Salt, ...) are very hard to use and master. You need focused and/or dedicated resources to do this. You can't make this a secondary priority for some new developer without significant risk.

Virtual Machine Setup Virtualization is key to all modern software deployment. Vendors have given us tools, such as Virtual Box and VMware, to build virtual machine images that guide the configuration of new servers. Except for specialized applications with special needs, apps are hosted on hardware that run many virtual machines simultaneously. This has produced huge cost savings for the entire industry and simplified operations.

Old applications were hosted directly on servers which meant that deployment required setting up the hardware and configuring it to support the application software that would run on it. This is now replaced with the challenge of automating the machine configuration for the virtual image. The VM image is then provisioned in the host environment so that the actual server can start serving the task it was created for.

Creating new VM images can either be done manually by starting with an operating system and building up the appropriate packages or by writing scripts that can automate this process. Using CM tools can make the process of configuring new VMs far more repeatable. It is also possible to write shell scripts that can be used to install and setup all of the server configuration that you need.

Reusing virtual machine images is a key benefit of virtualization. If you need to create twenty servers with the same software you just create one virtual machine image and instantiate it twenty times. This lets you create a template for every new server while letting the servers have their own life cycle and data.

Configurations So what is in a configuration when we are preparing to use a configuration management tool? There are several key areas that need to be configured to produce a new server instance.

- *settings* - Many different apps have settings that must be configured for the app to run properly. It may require details of the host name, network topology, or features to be enabled.

- *programs* - The software is typically a stack of programs rather than a single program running on the server. Each of these programs must be installed and configured properly. CM tools let you quickly install programs on top of the basic operating system services.

- *users* - Modern applications often have groups of users and the corresponding permissions granted to them. When creating a new server you may need an easy way to bring in your user data to the new machine.

- *network* - The network information is an important part of the application. You need to be able to assign network details such as domain name, host name, network IP info, and security settings based on the situation.

- *services* - Your application is most likely a set of web services that coordinate with each other. There is some configuration information that governs the behavior of the interaction and collaboration of these services.

Examples of CM Tools   *Puppet* https://puppetlabs.com

Puppet is an engine to help setup and configure machines and automate common deployment tasks. "With Puppet, you define the state of your IT infrastructure, and Puppet automatically enforces the desired state. Puppet automates every step of the software delivery process, from provisioning of physical and virtual machines to orchestration and reporting; from early-stage code development through testing, production release and updates."

Chef https://www.chef.io

"Chef is a company & configuration management tool written in Ruby and Erlang. It uses a pure-Ruby, domain-specific language (DSL) for writing system configuration"recipes". Chef is used to streamline the task of configuring and maintaining a company's servers, and can integrate with cloud-based platforms such as Rackspace, Internap, Amazon EC2, Google Cloud Platform, OpenStack, SoftLayer, and Microsoft Azure to automatically provision and configure new machines. Chef contains solutions for both small and large scale systems, with features and pricing for the respective ranges." https://en.wikipedia.org/wiki/Chef_(software)

Ansible http://www.ansible.com/

"You need a consistent, reliable, and secure way to manage the environment - but many solutions have gone way too far the other direction, actually adding complexity to an already complicated problem. You need a system that builds on existing concepts you already understand and doesn't require a large team of developers to maintain."

Salt http://saltstack.com

"SaltStack software orchestrates the build and ongoing management of any legacy or modern infrastructure. SaltStack is known as the fastest, most scalable systems and configuration management software for event-driven automation of CloudOps, ITOps and DevOps."

Lightweight Virtualization Using Docker   Recently there is a new paradigm for creating the execution environment for software. Lightweight virtualization has some significant advantages over traditional virtualization techniques. A virtual machine abstracts the hardware so that software written for any operating system can run.

This means that each virtual machine must carry the full operating system inside each VM. Lightweight virtualization is built on abstracting a set of software services so that software can be deployed in a much smaller container.

Benefits of Lightweight Virtualization

- Application containers are much smaller than virtual machines
- Containers load faster
- More containers can be hosted on one hardware server
- Every container holds all software dependencies with explicit versions
- Containers are portable between hosting providers
- All major public cloud providers are producing provisioning tools
- Software is built on a fixed standard operating system layer
- Tools allow you to go from development to production with the same execution environment

Many containers that are hosted on a machine will all share the same instance of the operating system kernel. This has huge impact on the number of containers that can be run on a given machine. The start up time for the service is greatly reduced since there is no need to load the kernel. The overall system load that a container produces is far less than a virtual machine.

_Docker is a platform for developers and system administrators to develop, ship, and run applications. Docker lets you quickly assemble applications from components and eliminates the friction that can come when shipping code. Docker lets you get your code tested and deployed into production as fast as possible.

Excerpt from http://docker.com_

One of the best new technologies for lightweight virtualization is Docker, which was introduced at PyCon in 2013. Since that time it has received a great deal of attention because it offers real answers to many pressing questions. I believe that Docker represents the future of software deployment. Time will tell, but Docker may emerge as the best answer for packaging, distributing, and running apps.

There are several core concepts that Docker uses. Here is a minimal introduction to the vocabulary. For more information visit the Docker website at http://docker.com.

- *Container* - A container provides a minimalistic version of a Linux operating system that gives your app a run-time execution context. Starting up a service process involves loading an image and starting the container.
- *Docker Image* - An image is software you load into a container. An image has all of the required dependencies configured to run the required service. Images can be stored for later execution or extension.
- *Docker File* - Images are constructed from other images by following the recipe within a Docker File. Therefore, the Docker File is like a build script for Images. It provides configuration instructions for an image instance.
- *Docker Hub* - A web service (like Git Hub) lets you find and share Docker Images from others. This creates an entire ecosystem for collaboration.

Containers bundle all the required dependencies for an app. Loading an image into a container sets up access to all of the dependencies. It should be as simple and small as possible to startup fast. Images are the template for instantiating new containers. Repositories allow you to share images with others and there is a public repository called Docker Hub. Docker file is the recipe to configure an image. Often new images are built on existing images that already have most of the dependencies that are needed. The developer only needs to add new requirements.

Provisioning Servers

Large applications require clusters of computers that are collaborating together. Orchestration is the task of managing all of the different servers that are running. As the load climbs you may need to provision new services of a specific type.

Tools automate the startup, shutdown, and monitoring of running servers. All of the major cloud service providers offer extensive tooling for managing your services. They support automatic rules for when to add new services and when to reduce the server pool. They also give you tools for monitoring many different aspects of your operation in real-time.

Hosting vendors typically provide tools specific to them. Spend time learning the tools that are related to your cloud provider. Experiment with the options for the specific scenarios that you are most interested in. As you scale your system load, the key bottlenecks are likely to move. Perform experiments to create scenarios that are typical of what you will encounter.

Clusters of Dependent Systems There are different types of servers that will be important in your operation. Make sure that you have solutions that are appropriate for the real challenges that your business faces. The architecture that controls these server interaction is also very important.

- *load balancers* - manage rules for routing the requests to other servers
- *web servers* - map HTTP requests into HTTP responses
- *application servers* - business logic for your app
- *databases* - relational or No SQL services provide persistence
- *caching* - remember and offer responses to speed up throughput

Each type of server may have multiple instances running. A cluster of servers may be used to handle all of the required traffic. Tools will let you manage the number of servers that are running and provide rules by which new servers are started or shutdown.

In a complex production environment rolling updates are typically done to deploy new software. There may be hundreds of servers running and it is not practical to upgrade all of them at once. What usually happens is new servers are brought online that contain the new software and are meant to coexist with the old system. The old and new software will both run for a while together, but eventually the old servers will be shutdown, leaving only the new software to handle the incoming requests.

System Setup with Provisioning Tools Leverage can be applied to how we manage operations. The established policies and processes can be reused and applied on each new release for a period of year. But for this to occur, our operations must have three characteristics. Automated, Dynamic, and Autonomous systems support the leverage you need.

- *Automated* - Situations must be automated and managed by scripts rather than by staff knowledge enacted manually.
- *Dynamic* - The automation must be dynamic and anticipate a broad range of new circumstances. Rules should dictate the resources applied to meet the existing demand.
- *Autonomous* - Systems should be autonomous, running independently until attention is needed to fix a problem.

Business rules should control the Computing Resources and Storage Resources that are applied to address the current load. A simple configuration setting should give you the control necessary to manage the fluctuating demand for your apps.

Another point of leverage that should be evaluated is the degree to which you need cloud portability. Are you locked into your existing cloud service provider? Do you have the flexibility to move between providers if needed? This is often dependent on the tools that you use to manage your operations. To minimize vendor lock-in you should lean more heavily upon tools and technology that support that provide portability.

Best Practice #10 - *Automate everything except the decision to deploy.*

Problem Operations frequently depend on the knowledge of specific staff members. The loss of a single individual can often threaten the entire operation. Processes that rely on people are not easily repeatable, and will limit the amount of leverage that is possible in your operations.

Virtual machines are typically used to deploy applications for customer use. It can be a significant task to get new VMs properly configured and provisioned. Without a proper set of servers and automated scripts, deployment can be messy and painful. All configuration information must be managed under version control because it is an integral part of the software.

Solution Using configuration management tools can drastically simplify the setup of new machines. These tools produce settings that can be committed to version control and applied with repeatable results.

Deployment should be practiced on staging and testing servers in preparation for a scheduled update to the production servers. The process of building and provisioning new servers must be managed by automatic scripts. Docker provides a new model of lightweight virtualization that makes it easy to move code from development into production. Large commitments are being made by public cloud providers to make it an effective solution for future deployment.

Next Steps

- Evaluate the number of clicks required to create new VM Images.
- Evaluate your process for managing active resources based on current load.
- Investigate Docker as a deployment alternative.
- Develop a block diagram of server types and how they are managed.

Chapter 11 - Monitoring Operations

"To improve is to change; to be perfect is to change often."

~ *Winston Churchill*

The Metrics Mindset

To control our applications in the wild we must gain a thorough understanding of how they behave. A system is designed to operate in a certain way but is it really doing what we expect? Monitoring is how we find out. Our desire is to get an accurate picture for how the application functions in live usage. This is different than static analysis which looks for issues in the code itself. Monitoring is about watching what is happening in the real world and detecting patterns that require attention.

Leverage requires a deep system understanding. Measuring things is the fastest way to gain that understanding. Observing the system in operation and extracting data can yield profound insights. It can also validate or refute our assumptions and beliefs. The scientific mindset enables us to make hypotheses and experiments in order to better study the system.

Our goal is to ensure that everything is running exactly as we intended. Even in a small system this will entail checking several hundred things continuously. Obviously this can't possibly be done manually so instead we invest in automation to monitor everything that we care about.

Adding telemetry to your applications is the first step. With high level languages or environments it is very simple and cheap to add an embedded web server. You can report point-in-time information such as loaded libraries, interesting configuration parameters, rates, and counts. This enables you to directly probe the application for its current state. If done in a standard way, it provides a basic way of enabling distributed monitoring. As you scale, you can replace this simple monitoring framework with more scalable capabilities built on something like Apache Kafka.

Test What You Care About Continuously Each tap point for monitoring can be built using the same patterns as the test cases that we built in the chapter on testing. We will use the diff test framework to check every test against the approved answer. This makes it extremely easy to write and maintain tests. Every test case should be one or two lines of code.

Here are some examples of the types of tests. Even without knowing anything about the problem domain or implementation language these scripts are easy to read and understand.

```
def check_disk_space():
    system ('df | ok_space')


def temp_files():
    system ('find | limit_lines 1800 2000')


def reports_filed():
    from reports import report_count
    num_reports = report_count()
    if num_reports < 100:
        print("Not enough reports "+str(num_reports))


def process_running():
    system("ps -ef | grep apache | line_count 2" )
```

Bad Events Should Trigger Automatic Escalation The diff testing framework will save many hours in constructing new tests. Instead of carefully thinking through what should happen, you can build test cases that generate output. Then you can decide if the current answer is acceptable. You can also let tests be extremely noisy at first and then silence them selectively to ignore irrelevant output.

This lets you spend effort on the things that need attention and ignore all of the things that produce the same stable output for every run. The effectiveness of this approach in the real world can't be overstated. Every test can be created in less than five minutes and repeated failures can also be addressed quickly. The only tricky part is transporting results to a general remote monitoring station. A simple solution for notification of escalation events can save a lot of time.

Every condition that needs attention should get it. No failing tests should be ignored. False positives are typically fixed by filtering the results of the monitor or broadening some constraints before the check is being done. A common design pattern is to make all tests only output exceptional behavior. All test cases would be written to remain completely silent in the healthy case and only generate output that requires attention.This is not required but may work well for you.

Often, exceptional behavior requires tracking trends and baselines. Tools can compare current results with expected results to look for patterns related to typical behavior. Appropriate behavior may be related to situational variables, such as time of day or user load.

In practice you will produce well over a thousand test cases. I recommend caching the results of long running tests. This allows you run the tests often without an undue burden being placed on the system. Only occasionally will the long running tests be executed. Most often the cached result will simply offer up the last answer. Note that failing test answers are not cached so that failing tests are run repeatedly until they pass.

Strive to have the monitored application report its cached "result". It is in the best position to manage that information. This works really well if the transport is cheap and telemetry is well designed. The burden of analyzing data is the responsibility of a third-party app. Collecting this sort of data every several minutes isn't that burdensome for the app but it is best to avoid designing complex monitoring stations that talk directly to your app.

Operational Failure Should Result In New Test Conditions No system is perfect - there will be situations that happen and are missed by the current test cases. Every time this occurs it indicates a testing weakness. Your testing boat has a hole in the hull. Patch it soon so that when the problem recurs you will detect it immediately.

Over time your system will become very robust. It will be constantly monitoring a large number of issues. Occasionally something will happen and an alarm will go off but most of the conditions will remain silent. When something does go wrong you will be the first to know. You will have an advanced early warning system that will tell you what you need to do and where you need to look first.

Do Not Ignore System Errors Or Warnings For your system to increase in quality you must respond to the issues that need attention. Robustness is built simply by fixing every problem that occurs and building additional monitoring each day. Monitoring is a lifestyle that prevents problems by addressing weaknesses before they cascade into something more serious. A small hole that is patched today can prevent a giant hole later.

Be sensitive to the smallest symptoms and don't be hasty in silencing complaints from your system monitors. Files created at the wrong time or transactions not processing may not be a problem but they may indicate an unexpected behavior that could be the first indication of a serious issue that will cause problems later. Respond to the pre-detection in order to prevent emergencies.

Instrumentation and Logging

Build a standardized logging facility to record each call to strategic functions within your code. Think through how to best use the logging system. What are the best ways to format your messages, provide app-specific message fields, and to delimit the data. Thinking about these things ahead of time will simplify your life tremendously and give you a head start if you eventually move to a log analysis and data mining tool like Splunk or Elasticsearch.

Think of your logging capability as video surveillance for the traffic. It allows you to find control and data flow issues when things go wrong. It also lets you see the sequence of events in healthy operation. Logging can give you tremendous insight into your software with minimal effort.

Be strategic about the tap points you select for logging. Each module has a primary entry point that is the gateway to its core functionality. This is an excellent point to place a sentinel. Log all of the traffic through this gate. Log the time, primary function, and key parameters. This will let you construct the story at the interface later.

Put Instrumentation in Each Layer   Your architecture is undoubtedly built in layers. There are operating system services that support the generic tools, frameworks, general purpose libraries that serve the business needs of your domain, and application services that meet the very specific needs of the current product.

Build separate logs for each layer of the entire system. Utilize 'logrotate' or a similar mechanism to manage when the current files are swapped out and how the older files are named. Build a mechanism to manage the data in such a way to easily reconstruct a history of past events.

Record All Interesting Data   The data that is recorded from the logger is typically time sensitive so most entries start with a date stamp. Another popular choice is to record counts of things. It is often useful to watch transaction frequencies and latencies and queue levels in real time. A simple stats class can be used to accumulate counts and then record the statistical information with each hit to the log.

By gathering individual sample info we can easily calculate the min, max, and average values of the set. This is very useful for building an understanding of the way the system operates. A simple bucket sort is useful for gathering the samples.

Unified Data Handling   Each time a function is called a log entry is generated. This entry can either be placed in a database table or appended to a file. For higher transaction levels more elaborate designs are required to batch up the data and efficiently write it to permanent storage. Start with something simple but adapt to the true business needs as they grow.

If you are logging data at 50,000 transactions per second you have completely different needs than at one transaction per second. Build your logging to match the actual demand levels. Logging is just like any other part of your system, the goal is to be as simple as possible while still addressing the fundamental demands.

Troubleshoot Complex Problems When problems occur, your logging system can be used to reveal the underlying cause. Special tools can be written to look through the raw log data to find patterns. By scanning the execution history you can identify the underlying sequence of events. The simple nature of the log entries makes it easy to construct tools to answer specific questions using regular expressions. Consider writing tools that have a state machine to detect sequences within your log files representing failure modes in your system.

The log data can be a treasure trove for analysis and experimentation. In more complex scenarios, commercial tools can be very useful. These tools record, analyze, and report on the logged data and contain detection for a wide variety of failure modes. They also are capable of providing great reports of the operational characteristics of your system.

Logger Load Because your logger is running as an integral part of every operation it is important that it be efficient. Be careful to avoid the Heisenberg problem, whereby the testing alters the behavior of the system. Make sure that your logging isn't placing an unwanted delay on every transaction. The performance impact of logging should be assessed early in the development cycle. Consider implementing logging as a closure, if your language supports it. This will allow you to be liberal with your logging while minimizing the cost of constructing the log lines if they aren't needed at the requested log level.

For simple monitoring needs I recommend starting with a simple custom solution. A few lines of logging code may be all that you ever need to support your overall project goals. As your needs get more demanding you can evolve this simple system into something more sophisticated. Eventually you might want to replace the entire system with a commercial system that offers high- efficiency recording, analysis, and reporting tools.

Analytics and Dashboards

Create a dashboard to make it easy for everyone involved with operations to see the high-level picture. A dashboard should be easy to read and give you an indicator of health within a few seconds. Many groups rely on personal conversations and reports but this approach is error prone and requires human effort. These conversations and hand-generated reports can be easily replaced by an automated dashboard. The cheapest dashboard can be a simple embedded web server serving up telemetry on demand.

Project Dashboard The dashboard doesn't need to be complex to be useful. For example, imagine a set of scripts that automatically generates the following text file on a shared server:

Issues	15 Open	100 Closed	
Source	20,888 Lines	300 Files	
Test	102 Tests	85% Coverage	23% source/test
Complexity	10200 Overall		
Complex Modules	90 Rubilator	67 Framistance	50 Security
Last Month	23/100 Issues	18000/270 Source	

The dashboard should be updated automatically. Make sure everyone has access to it and make it the focal point of all progress meetings. It is all too easy for discussions about progress to turn into anecdotes about what people are intending to do rather than a discussion of actual accomplishments. Inspire healthy competition within the team to "beat the numbers".

Allow everyone on the team to make suggestions for improvements. Respond to suggestions for improving the metrics. Often, other engineers and managers will have strong opinions about what you are measuring and how that data will affect their lives. Listen to their opinions and incorporate their ideas into each iteration.

Numbers are not a substitute for true understanding and judgment. Be sensitive to how the data is used and its human impact. Our goal is to build opportunity for everyone so this should be a motivating and energizing task for everyone.

Prefer quantitative data. Sound business decisions are supported by data that is gathered directly from the trenches. In the absence of real data, the strategic planners are likely to resort to wild assertions and crazy ideas. Build data that makes decisions easy.

Gather The Data Regularly All of the data gathering should be fully automated. Build scripts that gather all of the information at a reasonable time interval. If the data collection is cheap enough then it can be gathered frequently, while more expensive data can be gathered less frequently. Make sure the data collection is as automated as possible, otherwise it will be neglected at some point.

Scale requires you to flip monitoring on its head. In these situations applications push reports and monitoring stations subscribe to measurements that they are interested in tracking. Telemetry flows continuously through a bus and monitoring stations can then pick what is interesting.

Count many different things. Record latencies, throughputs, queue depths, number of records, and other types of volumes throughout the system. For each of these attributes, think of the healthy range that represents normal operation. Set thresholds for each attribute. Report the range violations as an early warning alert. These alerts can be used to identify problems before they have even happened.

Counts beyond the expected range can trigger escalation to operations. Exception-based Monitoring focuses attention on the critical areas. Pattern matching can also be used to watch a combination of conditions to provide even more sophisticated warnings. If you are watching a thousand different conditions in real time and ten are in a warning state then you can rest easy. 990 conditions are totally nominal allowing you to investigate the remaining ten. This will give you a deep understanding of your overall system. Just investigating the pre-warning alarms will let you control your system at a level that few operations ever achieve.

Design Dashboards Design your entire monitoring strategy to support your business. What do you need to control the most in your system? Build your monitoring around those needs by making sure they are represented in your dashboard. Make the key metrics highly visible to everyone. When the values start to creep outside the acceptable range alarms should go off.

If you are monitoring a thousand different things do not show a list of all of the things being monitored. Instead, use top-n reporting to show the items that currently need attention. Build different severity levels for different conditions. Allow people to monitor errors, warnings and informational attributes with different parts of the UI. Some people will only be interested in errors and there should be an escalation path that just shows that.

Notification System Design a notification system to provide a method for escalation of the most critical issues. Some issues demand immediate attention and must be sent to the responsible parties immediately. Email and text notifications should be reserved for critical issues and other issues can flow through the web dashboards.

You can easily build your own notification system using tools like Amazon Simple Email Service (SES) or Google App Engine Mail API. Or use commercial systems to provide the sending capability. SendGrid, Mailgun and Mailjet may be worth looking at for a general solution. You can also use tools like SNAP from C.A. Technologies to build your notification system. Seek to strike a balance between simplicity and meeting your specific business needs.

Measurements Drive Decisions

Metrics will drive behavioral change. You should spend some time researching the area of greatest need for change. Avoid random measurements that don't really matter to you. If you are starting a new metrics collection it is important that you start with the most critical things first.

Collect the data to quantify your core development process capabilities or the quality of the product itself. Metrics related to the development process and software product should be the initial focus. Collect the numbers that best illustrate what quality looks like.

Be careful not to sabotage your data collection. I've been involved in a number of data collection projects that met immediate resistance because it wasn't clear why data was being collected or how it would be used. Engineers are naturally suspicious about how data might be used against them so be sensitive.

Start with low-hanging fruit. What data do you already have access to? What data could you easily collect with a 20-line Python script? Would this data give you insight? If so, grab it and use it. Almost everyone has access to issue counts, line counts, tests passing, etc. Use what you have and continue to build more.

Baseline, Trend, and Goals Create a baseline by selecting a metric that determines what normal looks like. If you have previous data you may be able to leverage that knowledge to calibrate a baseline. If you don't have adequate data, it's not a problem. Once you start looking at any number consistently you will develop a feeling for good and bad.

Question the initial numbers and don't take anything for granted. Validate the measurements and ask questions about why the data looks the way it does. Remember that the true quest is for deeper understanding, not just pretty graphs. Propose a plan for possible improvements. Think about how th knowledge you've gained could be used to take your project to the next level. Metrics aren't the final answer, as much as they are a source of new questions.

Validate your approach with others. This is a chance for you to engage the rest of the organization in a larger discussion about software development. Show others your early data and gather their opinions about where you should look next. Allow collaborators to influence your direction and contribute ideas to the project.

Experimentation and Quantifying Results The goal of experimentation should be to figure out what works and what doesn't. The metrics give us a feedback loop. Determine how to best test the affects of possible changes. Design an experiment to test a hypothesis or validate an assumption.

After the test is in place apply a desired change and study its impact. For example, we may believe that high complexity lowers productivity. This requires measuring both complexity and productivity. It may take several iterations to get a good complexity measurement or a productivity metric. As a result of this work, you will have a far deeper understanding of, not only complexity and productivity, but the interactions between the two.

Next, establish operational rules that are customized to the needs of your organization. Commit to a process of positive changes that are driven by empirical evidence. Step back and take a longer view by examining the bigger picture for your development projects. What are your biggest unmet business needs? Where are you able to excel in areas where other groups struggle? When do you stumble and how can you improve to have better results? Are you moving in the direction you want? Think about where you will be a year from now. Where were you a year ago?

The software industry is rapidly advancing and changing. There is a coming labor shortage that will catch many companies by surprise. What does the actual data teach you about your organization? Will you be able to attract top talent in an environment where engineers can have their pick of places to work?

Companies that build a robust development capability will produce far more and have an attractive work environment. Measurement can be at the heart of your ability to build a world-class software company. Organizations that run repeated death marches due to poor planning will fail in the next decade because they will be unable to staff their projects.

Many organizations measure random things without a clear view of what they are trying to accomplish. This is a lot like engineers making random changes to a complex system hoping to fix a bug without really understanding how it all works. In both cases, we need to disassemble the system to examine each part in isolation.

Start with the overall goal, then figure out which measurable attributes you can observe. Understanding the fundamentals will help you optimize them and design a path for improvements. Steer the larger effort by conducting small experiments to validate your assumptions.

Testing is the key to software process improvement. If you can think of an idea and a way to measure the results, you can validate that you are making positive improvements. Testing also gives you evidence that your improvement is real. Then you can lobby for broader adoption with confidence.

Drive Business Decisions Some organizations collect data without a clear plan of how it will be used. Collecting metrics is really only valuable if you use the data to change how you do business. Metrics should always be used as part of a bigger plan to improve your software processes. Make reviewing the metrics a key part of any business review. Find the most compelling numbers and concentrate on them.

Be selective about the metrics you gather. Gathering a few key numbers should be more than sufficient to answer the fundamental questions. Communicate the decisions in terms of the measured results. The focus of measurement is improvement, enabled by better understanding.

Essential Project Metrics There is no standard set of metrics that will work for every organization so the goal is to calibrate the metrics for your development team. What works for one group will be a danger signal for another. Figure out the most critical measures for your team at this moment. Starting with too many metrics will generate confusion and undermine the overall effort.

Once you have selected the values to record, you need to figure out how to collect the numbers automatically. Avoid any metrics that require a person to record data. A skilled Python programmer can write a tool to collect almost anything in about an hour.

The next step will be to calibrate the baseline. If you are just starting out collecting data, I recommend that you defer creating any basic rules until you have been collecting values for a while. Start by measuring the current reality and then apply judgment later.

The absolute values mean little but the trends are important. Until you have collected a lot of data over time you can't speculate that 100 defects or 20 tests is good or bad. Remember, you are looking for changes to understand more about your ability to successfully create software.

Abandon metrics that don't seem to yield actionable insight and expand the metrics that you find useful. Create finer-grained resolution on the quantitative data. Create a competency related to collecting and using data to foster improvement. The following lists gives an overview of possible metrics. You may want to start by selecting a subset of these items to measure.

Issues

Problems surface during the course of a project and should be logged so that they can be tracked by the team. The issue tracking system can be mined for relevant information about the status of the project and what areas need the most attention.

Consider developing a simple set of scripts that interrogate your issue tracking system and supply counters on specific areas. These types of tools are very easy to build and access to the tool implementation makes it simple to adjust to the changing needs for information throughout the course of your project. Here's a sample list of issues to measure:

- Features to be implemented
- Features to be defined
- Architecture and design issues
- Failing tests and defects
- Defect finding rate - Hours of testing to find the next defect
- Defect fix rate - Useful for projected completion
- Defects deferred

Code size

Code size is an important indicator of the areas where the team is spending the most time on the project. Just by looking at the size of the code and the different types of code will reveal a wealth of information that is useful for managing the project.

For many languages, very sophisticated tools exist and are freely available. You may want to investigate sonarqube.org. This tool counts the number of source code lines in different areas of your product. The exact information is not important but the comparisons are extremely useful.

- Modules
- Functions
- Lines
- Documentation
- Tests

Compare these measurements to each other. Ask key business questions and let the data reveal the answers:

- What percentage of the code is tests?

- What percentage of the code is documentation?
- How many functions per module?
- How long are my functions?
- How long are my tests?

As you begin to take measurements you will naturally begin to form ideas about the proper rules for ideal code. Instead of being a pre-determined opinion about the correct basic rules, it will be a well-informed opinion based on data.

Run the tools each day to ensure that the code is changing in a way that is consistent with a healthy project direction. By creating plans that are tied to actual code measurements you are leveraging the knowledge of your most experienced engineers. You are also training your junior staff to rise to that level.

Complexity

Another important area for measurement is code complexity. As we discussed in chapter 6, Code Leverage, measuring the complexity of various parts of your system is critical. This information needs to be fed back into the planning process to generate the standards of complexity for a release criteria. If you don't decide on this ahead of time the pressure to ship will undoubtedly overwhelm the quality argument. This leads to releasing code that has an unacceptable level of complexity and poor quality and will eventually turn into a maintenance nightmare.

There are several different ways to study complexity. Create tools that capture readings automatically and follow up on the advice they provide. Each of these techniques provides a unique viewpoint into your system. Use the ones that make the most sense to you.

- Points of interconnect
- Halstead & McCabe complexity measures
- Seaman's Complexity sum(lines ** 1.2)
- API language vocabulary
- Code Stability (%code change)

Team Productivity

A simple measure of productivity will allow you to see the impact of events outside of your project. It is easy to put some simple measures in place. These metrics can provide the key information you need when proposing project level changes.

- Velocity (features / engineering day)
- Lines / day
- % of features / day
- Compare this project to similar projects

Test Coverage

We all believe that testing is important but it's good to compute the percentage of effort related to tests compared to the product design. A product where testing accounts for less than 10% of the code is in serious danger. Another thing to watch at the project level is how many defects are repeat offenders. This may indicate a failure in the regression testing methodology.

- lines of product / lines of test
- recidivism (repeated defects)

Monitoring Tools

Measure all of the critical attributes of your system:

- Transactions/second
- Memory footprints
- Cache page-faults
- Max load conditions
- Concurrent users
- Disk space requirements
- Scalability requirements
- Developer productivity

Make a list of everything that is important to your application and take measurements. This becomes the baseline for adequate performance of the system. Any new system should create a demonstrable improvement on a number of these metrics. Serious scrutiny should be given to any measure that drops. A hard look is especially required before committing to any new technology.

Best Practice #11 - *Monitor everything that you care about.*

Problem Monitoring can produce a deep understanding of the operation of your software. It can yield insight into which parts can be leveraged and how to get the results you want. It can also teach you how to improve your development and operations processes.

Monitoring covers a broad range of techniques that include: instrumentation, logging, metrics collection, notifications, and dashboards. Together these techniques build your system understanding and allow you to make strategic improvements to your overall project. Neglecting monitoring will cause you to rely on assumptions and folklore that will often be incorrect.

Solution Monitoring begins with systematically instrumenting your code. Log files can reveal the history of your system at run time and let you write tools to analyze this history. Logic can be added to your application to collect key metrics. Dashboards and notifications call attention to key problem areas.

Each of these techniques will contribute to the quality of your products and services by lowering the technical debt. These best practices in software engineering will preserve the value of your investment for years to come.

Next Steps

- Implement a consistent policy for logging in your source code.
- Select five key project metrics to begin measuring.
- Plan three experiments using measurements to validate assumptions.
- Research monitoring tools that are available from you hosting provider.

- Begin developing standard rules about quantitative values for your key project metrics.

PART 4 - Culture of Leverage

Figure 12:

Chapter 12 - Knowledge Management

> "To know what you know and what you do not know, that is true knowledge".
>
> ~ *Confucius*

Leverage Understanding

Building software is about solving problems and requires a broad scope of understanding. Valuable understanding was acquired in previous work and can now be applied to new problems. Re-learning everything from scratch with each new project is extravagantly wasteful but that is what most groups do if there is no standard process for capturing and organizing knowledge. Throughout my career I have worked with many different development teams and very few organizations have had effective tools for managing knowledge. They may be very advanced in how they deal with source code but completely inattentive when dealing with knowledge that is not software.

A system and a set of tools is required to leverage our understanding of current problems into the solutions needed tomorrow. This type of leverage is often missing because we don't have an effective way to gather and share our team's knowledge. Building an effective way to deal with information is a critically important aspect of becoming more productive.

Many organizations default to relying on the memory of individual staff members or private notes on their computers. This is obviously unreliable and will lead to a lot of wasted time and effort to relearn valuable knowledge. Without a knowledge management system, vital information will be lost every time there is a personnel change on a project.

A storehouse of knowledge that truly captures everything that your team needs to know is essential. Create an easy way to find all of the information quickly and make it easy for people to contribute new information. If it is too difficult to add information it will cause people to ignore the system.

Now let's explore how to build a great information retrieval and sharing system that is low cost to set up and to run. There are several different aspects of dealing with information to consider. Understanding goes through distinct stages of refinement and different activities are needed to deal with knowledge throughout each stage. Think of these aspects as a "Knowledge Life Cycle".

- Capture - The process of recording new ideas as they occur
- Organize - Ideas need collaboration and connections to other related ideas
- Refine - Editing and critique are essential parts of building great ideas
- Share - To be useful, an idea should be shared with others outside the team

As knowledge develops through each stage the idea will be strengthened and refined into something that provides unique business value. If an idea fails to be developed in one of these stages it will tend to atrophy and die. Review is a key goal of having a well-defined process for managing information. By making it easy to see the progress of many ideas, you can also spot ideas that are stuck and in need of attention.

Capture - Where Ideas Go to Thrive

No idea is birthed fully-formed. A small idea may come in the form of a hunch or an unsolved problem so it's critical to capture the concept before it evaporates. A quick note is adequate to record the concept and spark the memory at a later time. Then this reminder can be used to more fully reconstruct the original thought. Ideas that start as a simple concept can be expanded into a full-blown product plan over time. Capturing information should involve the following elements.

- Identify the problem
- Identify the solution
- Identify the context where the idea applies
- Identify the limitations and caveats for applying the idea

Building a robust system to manage knowledge is easily done with tools that are low cost and readily available. My favorite tool is Evernote because it is easy to use and offers a broad range of services. You can also use a wiki that is either in a private or public cloud. The critical thing is that you select a tool that can be set up quickly, is easy to use, and one that you can commit to for the long term. Switching systems can be expensive and you want to avoid this if possible.

A knowledge system must be exceedingly easy to use - this is a critical requirement. Engineers are busy and they will only use tools that they see immediate value in. Anything that looks like busywork will be neglected and fall quickly into disuse.

Ideas come and go quickly - your tools need to be setup to take advantage of the fleeting nature of ideas. The system you define needs be accessible from anywhere. Whether the device at hand is a phone, post-it note, email, random text file, photo, or a screen shot - they must all be compatible with your knowledge management system in order to get people to accept and use it.

Window of Opportunity Any new idea must be recorded within 15 seconds because if it takes any longer people will skip using it, resulting in lost ideas. They will make a mental note and then promptly forget to follow through. Count the number of steps that are required to record an idea. Here is an example capture path:

- Light bulb goes on (attention: great idea generated)
- Grab my phone
- Click on Idea app
- Type a few words to record the thought for later
- Put phone in pocket

There is a narrow window of opportunity in which to capture ideas. When lightning strikes there needs to be immediate access to a means of recording the idea while it is still fresh. Ideas evaporate quickly when they aren't captured, but the very act of capturing an idea will spark other related thoughts.

Most engineers are teaming with ideas. The ability to collect the ideas and then have a standard process for developing them is vital. Capturing and processing ideas creates an incubator effect as ideas that start off as half-baked hunches can be cultivated to produce real gems. Your goal should be to provide a system with a smooth path from hunch to breakthrough.

Zero Overhead Capture In order to capture as many ideas as possible it is necessary to remove all of the barriers to record a thought. If your knowledge capture process requires getting on a computer and browsing to a document that can be edited you may have already lost them. Instead, think about a low tech solution as simple as a pen and a small pad of paper. Capturing the seed of an idea may be as simple as writing a single word on paper.

Now imagine replacing the paper pad with Evernote or other similar tools. Whether you are at your computer or just have your phone, you will have access to the same set of notes from anywhere. Evernote is extremely convenient to capture a single idea since you can also make short voice recordings or take photos as notes. Then when you're at a computer you can process these notes to make sense of the original intent. Provide several different on-ramps to make it convenient to record ideas in many different settings.

Universal Capture Capturing knowledge is a prerequisite to managing information so capture the full knowledge of each team member. Treat other types of knowledge with the same level of care that you use to track software changes because the non-software knowledge is often just as important as the software we so diligently protect.

Imagine setting up a system to track everything. Each discussion about project schedule, user requirements, design challenges, and alternative patterns that should be evaluated, is recorded.

There is power in teamwork and collaboration in this endeavor. Many brains thinking about problems will always produce better answers than people working in isolation. You can get creative to boost the collaboration process by doing things like running contests to see who can come up with the highest number of best answers to sticky design problems.

By having a repository of requirements, designs, problems, solutions, and stories, you have the foundation of widespread and meaningful collaboration within your team.

The goal is to record as much as possible so content can spur additional thinking and conversation. Avoid heavy requirements because that's a collaboration killer. Record many different kinds of information in your system. Here's some examples of things you should include:

- Formal design specifications
- Formal product requirements
- Product ideas
- Product road map for several generations
- Rejected requirements
- Requirements being considered
- Design patterns
- Coding patterns
- Coding standards
- How to set up tools
- Guidelines for version control
- Release guidelines and version numbering
- Systemic problems that need solutions
- Competitive analysis of similar products
- Project planning schedules
- Risk assessment and mitigation plan
- Contract templates
- Database schema
- One-page architecture diagrams
- Presentations on market research
- Presentations on project progress
- Product backlog of future features
- Test strategy and plans
- Tricky test cases
- Build tools
- Archival requirements
- Product support plan
- Warranty data from previous products

- Productivity metrics
- Retrospective summaries
- Customer use cases, customer needs (unsolved problems)
- Cheat sheets on tool usage

You get the idea - the sky is the limit. There are thousands of pieces of useful data that you should be able to find easily. Often they are stored in a file system hierarchy and it's very difficult to find what you need. Use your information management system to track the URLs to the complex binary data, but make it easy to find what people need. Engineers should never have to browse through a lengthy file tree looking for stuff. Instead, let them quickly find the content by querying for keywords.

Information may exist in many different forms in many different systems - this makes it very difficult to find and modify. Consider the advantages of bringing all this information together. Try to standardize the data format as much as possible. Moving to a text file representation makes the data universally searchable. Begin moving data from various locations where it currently exists into a single location. Then the knowledge management system can function as a resource for all development.

Build a system that is customized to your needs but use existing tools to solve all of the generic problems. You could build the entire system from scratch if you believe it is absolutely necessary. However, your time is probably better spent building other parts of your system. Realistically, you should be able to create something useful in about a day. After that you can justify the cost of making it better by the immediate benefit it produces.

Intellectual Property Software is all about inventing things that solve problems in the real world. This often generates ideas that should be legally protected. When resolving legal disputes over ideas the timing of the idea first being conceived and reduced to practice is extremely important. Sometimes this is the deciding factor in court cases that happen years later.

You should begin thinking about legal protection at the time the ideas are first conceived. Make sure that whatever system you are using for recording knowledge has a history that can be used in any future disputes. It may also be useful to have a regular process of archiving the repository in a safe location.

Protecting your intellectual property is a business decision, but make sure that the tools you are using support any decision that might be made. Legal protection follows the following general pattern:

- File an invention disclosure with a patent attorney. This starts the clock and marks the time of the original idea.
- Create and file a patent application with the patent offices where you would like the protection.
- The application will be reviewed and the patent office may ask for amendments to the application before it can be accepted.
- Once the application is granted then it can be used to protect the invention.

Your system to track and manage information should be consistent with the business goals that you have. These goals may change over time, so it is important to give yourself the maximum level of flexibility. The key attributes you should have are the ability to find information quickly and have a legally binding time stamp associated with the data.

Organize - Identify the Natural Connections

Once information in captured, it must be organized. One approach is to organize the knowledge as a series of articles because the topics covered by the various articles are naturally connected to each other. The information management system should use these connections. Organizing the data is about finding the connections between topics and allowing people to use them to find the related articles.

There are several different ways to classify information. The most popular method is to organize topics within a hierarchy. Computer file systems have made this the standard mechanism that is familiar to everyone. However, this is the least useful way to organize lots of topics and it has many severe limitations. The most significant problem with hierarchies is that they assume one strong paradigm for organization. Each topic is represented at one location within the hierarchy. Most real world problems don't fit into this model.

Another model is to organize topics using keywords. This has the advantage that the same topic may be classified many different ways. Applying keywords as tags can be done easily in almost every popular tool. Whether you are building your own tool from scratch, using a complete system, or building a tool using Evernote or similar app, using tags is the best way to represent topics.

The task of organizing topics involves finding the underlying connections and classifying the topics so they can be found later with a single query. Organizing your content should involve the following elements:

- Find connections with other related topics
- Ask how others solve similar problems?
- Identify alternative solutions
- Build out the concept into a complete article on the topic
- Think about how this topic will be queried later

Think of tags that can be used to describe the new idea. Search the system for similar ideas that may shed some light on how you might move forward. Search the internet for related topics. Capture interesting articles and incorporate them into your knowledge management system. Retain information about where new source material is found so that you can avoid legal issues by improperly using any intellectual property.

Connect Ideas don't exist in isolation, they are always connected. Finding the appropriate connections is an important part of the development. Using tags will classify the content into major groupings. This may be a designation of the format of the info, or the phase of the life cycle, or the technology that is used. There are many ways to classify materials and our goal is to use all of them. Next, explore the connections that you find through this classification by browsing through related topics and looking for interesting similarities.

Use hyperlinks to add hard connections between topics - linking related topics allows others to make these same associations later. This can often add critical insight. You can add links to articles about the subject or keywords to help someone search for related topics.

Unsolved mysteries and unresolved problems can plague systems for a long time. One of the most useful purposes of a knowledge management system is to highlight ongoing weaknesses. This is less formal and more flexible than an issue tracking system which deals with concrete flaws in the system. Your group *brain* can be a place to explore alternative designs in order to understand several competing ways to solve a particular problem.

Complete the Proposal Part of the organization stage is to develop the details of the idea by fleshing out the particular problem and propose multiple solutions. Next, evaluate related problems and solutions and select the best alternatives to move forward with the idea. Since no piece of knowledge lives in isolation, every best practice, tutorial, machine configuration tips, or feature list is connected to other bits of information. These connections are important to build into the description of the idea because they guide others in how to be successful when working with this topic.

After ideas have been captured they must be nurtured and cultivated. This entire phase is about gathering info and connecting it together into some kind of network. Synergy occurs as various pieces of information are assembled into a new picture of truth. New insights and levels of understanding may come abruptly as a flash or they may be the result of persistently working a slow hunch.

Substantial breakthroughs may take years to cultivate. People may come and go during that time-frame. It is imperative that you don't lose the information you are building over time. The best way to protect ideas is to create a system that truly embodies the working knowledge of your staff.

I was once on a team that had spent years cultivating a deep base of knowledge in a particular domain but the business decided to outsource all future software development. Over the next year about 90% of the knowledge was lost in the transition. This ultimately resulted in a massive level of waste and a dramatic decrease in the success of the next products. This loss could have been mitigated by having an appropriate system of knowledge management in place.

Managing Keywords Create tools that expose the data in a natural way. A vocabulary of 100 unique keywords can be used as tags to get any type of information that you wish to retrieve in less than a minute. Build your own Software Brain on top of an app like Evernote simply by setting up some tags.

My own personal Brain in Evernote houses about 1600 notes and is supported by less than a hundred keys. I search for one or more tags and can scroll through all of the matching items in chronological order. Finding a needle in a haystack has never been easier. Here are some example tags:

- design
- plan
- issue
- todo
- done
- code
- pattern
- test
- idea
- requirement
- presentation
- book
- image
- diagram
- money
- staff
- skill
- dependency
- metric
- task

If you want to get fancier you can add structure to your keys to help people with the selection of keywords. Here is a set of structured keys that could be used as an example when building your own.

```
Languages:
    Java Script
    Python
    C++
    C#
    Java
    Ruby
    Go
    HTML
    CSS
    XML

Operating Systems and Servers:
    Linux
    Windows
    Mac
    AWS
    Google App Engine

Development:
    Requirements
    Specifications
    Design
    Code snippets

Test:
    Test plans
    Test strategies
    Test tools
    Test cases
    Regression
    Coverage reports

Project planning:
    Progress tracking
    Metrics
```

```
Schedule
Slides for Management Summit

Tools:
    Tool stack
    Tutorials
    Internet guides
```

These lists will vary widely based on your domain. The primary lesson here is that you can make the list match precisely the topics that you wish to organize. How will people ask for topics? Give them those keys. Let the tags grow organically to reflect the underlying content.

Prioritize Not all ideas are worth pursuing. By capturing information in one place it makes it easy to select the strongest ones to pursue. Use a sieve approach to select the ideas that have the most value. Old ideas will sink to the bottom so it's unnecessary to manually remove them from the system. On the other hand, ideas that are being updated will naturally rise to the top of the priority list.

Prioritization is an important aspect of organization. Ideas that appear feeble at first may be strengthened as other people develop them. The most important areas of discussion and design will get extra mindshare. In this way, the knowledge management system can be a focal point for nurturing a healthy discussion of design and software techniques and skills.

Core Tool for the Whole Team If knowledge is to be truly shared, it must be viewed as a team resource. Bring the entire team together to figure out how to get the degree of collaboration you need. We have all seen plenty of examples of someone trying to push their pet tool. Without grassroots support for sharing knowledge you won't succeed in building a collaborative system.

Start small. Select a pressing problem and show how a simple tool could improve communication. As more people begin to engage, encourage their efforts. In the beginning, focus on specific problems where the system makes it much easier for everyone to be successful. Eventually you may want to assign a champion to carry the vision of knowledge sharing.

Add ideas that you gather on the internet as articles in the system. Capture content from other teams and use all of this to craft a rich collection of knowledge. Allow the repository to grow organically. The information should cover the range from highly organized to completely disorganized. This range helps the tool see more usage over a wide range of purposes.

Problem and Solution Pattern One distinct pattern that is very useful in a knowledge management system is a recipe. A cookbook is filled with recipes that teach the novice cook how to get an outcome that is designed by a master chef. This approach can be used to great effect here. A recipe addresses a specific problem and outlines a specific solution that can be applied to get a good outcome.

There are several key aspects about a recipe. It is simple and tactical and it addresses one outcome and approach. Later it may be expanded into a tutorial that addresses a range of side issues. A recipe sharpens the focus to improve the chances of success.

Key limitations should be highlighted. It must be clear when the recipe doesn't apply and there should also be information about the expected ROI and key success indicators. The recipe should cover the essential elements that need to be covered and then stop. A catalog of these recipes can be invaluable to an organization because it can really bring the team together.

Refine - Ready for the Sharks?

During the refinement stage, the topic is applied in the real world. A best practice is created to show others how to get good results. It may require some experimentation to find the techniques that work best. At this stage you will also discover the key limitations of the practice and how to avoid them. Peer review can help to polish the information and make it more useful to others. This stage is about proving usefulness and should include the following aspects:

- Evaluate competing alternatives
- Foster experimentation
- Identify limitations
- Tune critical parameters and attributes
- Review by others to improve the ideas

An idea isn't really ready until is has been scrutinized by others. Everything sounds great on paper but the real test comes when other people review the work and are allowed to make their own contribution to it. Seldom is a single designer so brilliant that they create ideas that can't be improved. The more an idea is reviewed the better it gets.

The system you set up can have a big influence on how ideas get reviewed and refined over time. The primary goal of this system should be to invite as much participation as possible. The ideas should be viewable, searchable, and easily connected with other topics.

A version control system must be used within your knowledge management system in order to prevent accidental deletions or corruption of the system. Some organizations develop complex rules about who can edit what content but this is almost always a mistake. If abuse and disagreement occurs it should be addressed with the people in conflict, rather than adding new constraints to using the system.

The greatest danger for any group is that information becomes isolated. If there is conflict it can be worked out. In fact, constructive conflict is a requirement for good design. If no dissenting opinions are ever heard then groupthink has taken over and innovation is dead. Use your info system as a focal point for design discussions and collaboration. Invite everyone to edit and interact.

You can use mistakes as an opportunity for training. Teach new engineers how they can get their ideas accepted without damaging the work of others. The system that houses the knowledge of an organization is a great place for public debate.

Review Reviewing and refining is a critical key to successfully maintaining information over time. "Write-only" documents offer very little value. A document must be read in order to be useful. It should be easy for someone to offer contributions as they are reading and it must be a natural and seamless activity. Systems that make it hard to collaborate will quickly fall into disuse. As articles are edited by multiple authors you create a vested interest in protecting and developing the content.

Once there are several interested parties working on a topic they can have an ongoing discussion. They may also choose to set aside time for periodic reviews. This is especially useful on controversial designs and requirements. Remember that constructive conflict is good and should be encouraged.

Experiment Some ideas require more than just discussion. Sometimes prototypes must be built to validate the concepts. The knowledge management system can be a good place to explore which designs and tools need further development. The key assumptions can be enumerated and the validating evidence presented. It can also be a great way to publicize the results of the experiments.

Many designs require tuning in order to function correctly. The authors can perform the work to find key parameters and post the information that will make it easy to duplicate the same results without having to repeat the experimentation.

As designs are shared throughout the organization it is useful to have a place to discuss the experience. If someone tries to use a pattern and fails, that is very valuable information. Why did it fail? How can the design be adapted to work over a wider range? Sharing experience is the sign that collaboration is truly taking hold in a team. Gathering a team history can make it easy to bring new people up to speed.

Test Assumptions In the book Lean Startup, Eric Gries discusses the key elements of building an agile startup business. One of the interesting principles in the book is that everyone has assumptions that we aren't aware of. He advocates testing these assumptions as early as possible in a project. The same is true of a software project - all of our plans and designs are based on assumptions and frequently these assumptions are unspoken.

The first step to validating our assumptions is to make them visible. Once we have articulated the assumptions we can figure out ways to validate them. The system we've built to manage knowledge is a great place to have this conversation.

Identify the key assumptions that will affect your business and your engineering work. How can you design an experiment that will let you get an answer to the pressing questions early? Identifying the key assumptions and testing them can go a long way towards addressing controversy on your project. Begin to put systems in place that will help you answer your most critical questions early.

Quantify Your Content As more of your information is built into the knowledge management system you have a great opportunity for data mining. Simply counting relevant types of information will give you insight into what you are tracking. How much of your content is related to project status, design, requirements, code, or tests? Is your information primarily tactical or strategic? These are important questions that can be answered quickly with a few queries.

There are a few metrics that are important to track. How frequently are new topics being added and existing ones modified? If topics aren't being added this system won't really provide any useful value. On the other hand, if topics aren't being modified it may indicate that no one is reading the topics. A topic that is read will frequently be edited to update the information. You need both of these numbers to be strong in order to have a healthy knowledge manager tool.

It is a good idea to periodically review the oldest content in the system because it represents the stalest content that is still being tracked. Some of the content is probably no longer useful at all and should be removed while other content may have historical value and should remain. It is a good idea to spend an hour once a month browsing through the articles to see if the system is moving in the right direction.

Share - Time to Go Public

As ideas are developed and accepted within a team they reach a point where they are ready to share more broadly. Many problems are commonly seen in different places. Select the problems that you think might have the greatest impact for others and evaluate your solution. Have you solved the problem in a sufficiently general way so that it is useful to others, or is more work required before it is ready?

The sharing stage involves these important tasks:

- Be Crowd-Sourced
- Provide opportunity to vote for the best ideas
- Standardize the application
- Identify future steps

When you have a great solution available, share the good news with others. Build a prototype to show how the idea can be used in a context beyond the original usage. Create a one-hour training seminar to entice others to use your design. The effort invested into inventing something cool should be leveraged as widely as possible.

Look beyond the basic reuse of the code library to a deeper level. Look for design idioms and best practices that could also be shared with others. Code reuse is great but there are additional ways to leverage your new understanding. Build a great solution for your product *and* keep looking for ways to get even more value through leverage.

Showcase A prototype is the best way to get your idea accepted by others. If you build a solution within the context of your main product structure then engineers on other teams will be reluctant to accept it. There is some unknown amount of work to isolate the breakthrough idea from the natural entanglements of the product code. This work has to be done in order for others to adopt your concept. Build an isolated prototype that is as simple as possible. It should showcase your contribution and stop there.

Once you have a strong prototype that demonstrates the relevant concept, turn it into a great presentation with drama and humor. Build a slide deck that culminates with a live demo. Demonstrate the benefits of the design to anyone that will listen.

The knowledge management system should be used to provide all of the relevant documents that will support your entire concept. Build a single "Getting Started With ..." document to provide everything needed in less than an hour. Provide a User Guide, Installation Guide, Concept Proposal, Design Info, and source code for the prototype and tests. Packaging makes a difference so don't skimp here.

Contests Imagine a culture of sharing, learning, and inventing. Leverage thrives in an environment rich with collaboration. One great way to promote this type of leverage is by sponsoring an Inventor's Contest. Challenge the participants to show off the coolest thing they have invented in the last year. After people become familiar with the event it could easily be many individual's favorite day at work.

Capture these ideas and all of the artifacts in the info manager. Remember, our goal is to save everything so that we can get the maximum leverage from every document. This will become a treasure trove over time and a source of ideas for discovering new solutions to problems.

Beyond This Team Why move your sharing beyond the borders of your own team? As software architects, we have responsibility to the entire organization. Leverage the success that you have within your local team to influence others. Be a leader by focusing your energy on the benefit for the wider community.

By considering the needs of other parts of your organization you will certainly discover solutions that your local work group needs. Capturing feedback from others will influence your team in a positive direction. The nature of software architecture is that it rests upon a broad understanding of problems and solutions. Your interactions with others will broaden and deepen your understanding and make you a better architect.

Learn from Others You aren't the first one to solve problems. Others have already been there and done that. Make every effort to learn what has already been done then build new solutions where the existing ones break down. Make fundamental contributions to the designs of others and breathe new life into them. Copy great solutions and make them even better. Evaluate, improve, and share designs on a weekly basis. One of the key differences between architects and other engineers is their ability to look beyond the specific problem to provide a broader and more general solution. This is the heart of leverage.

Ideas benefit from interaction. Ultimately your success at leveraging the knowledge you have will be largely determined by your ability to gather and organize information from a wide variety of sources. Collecting documents is one thing, but your goal is far more ambitious. The real goal must be to make sense of all the knowledge that we have. Raw data must be turned into real, actionable information.

Best Practice #12 - *Create a robust system for sharing all types of knowledge within your team.*

Problem Many organizations have weak methods of tracking vital information. The actual source code is managed effectively while all of the non-source information is in danger of being lost. It is very expensive to relearn lost knowledge at a later date because it often comes at a critical time in the project which deepens the crisis even further.

Information needs to go through a series of development stages that are similar to the software life cycle. The organization needs a way to track the knowledge throughout its development and find key topics quickly when they are required. Without a shared knowledge management system in place, few of these requirements can be met. This often results in extravagant waste.

Solution A knowledge system is required to manage all of the different types of information that are required by the software team and, possibly, the larger organization. A custom tool can be built from scratch or assembled from other pieces of software (for example Evernote). The system should operate to address the needs of the knowledge life cycle:

- Capture
- Organize
- Refine
- Share

Each distinct stage of idea development has its own challenges. The information management tool provides a complete solution to all of the different aspects of managing knowledge. This system provides a nexus for sharing ideas and team collaboration.

Next Steps

- Make an inventory of knowledge that you need on your project and record its current location.
- Think about the development stages and what information is at what stages.
- Decide on the platform that can be used to create your information management system.
- Build a simple repository
- Create a migration strategy for bringing in existing data.
- Measure use and react

Chapter 13 - Teamwork

"Talent wins games, but teamwork and intelligence wins championships."

~ *Michael Jordan*

Build a Culture of Leverage

Products reflect the organizations that create them. Every organization has a personality and a culture that will dictate what the human experience of working there will be like. Some organizations are filled with life, they inspire each person to bring their best each day. They encourage risk-taking and exploration even when it may lead to failure. They build teams that freely collaborate with each other and provide an environment where everyone has the freedom to succeed. They create products that delight customers and are just right for the marketplace.

Other organizations focus on heavy control. The leaders at the top make every decision and the staff is expected to carry out these mandates without question. There is a lot of talk about loyalty and there are rules for everything. Fear is pervasive and failures are met with punishment. This is the old "Command and Control" management style that works somewhat in low- skill environments but is disastrous in creative companies that live on innovation and new ideas. An organization running on a strict hierarchical system will create products that are confusing and difficult to use. A persistent decline of market share is the inevitable result of the dysfunctional environment within the company.

Well-run companies will naturally produce high-quality products and it is possible for a good team within a dysfunctional company to have a history of repeatable successes if they are isolated enough within their company. Great products come from great teams. Of all the things that an organization can do to improve its software development capability, building great teams will create the largest benefit. Building high-performance teams is essential to every other aspect of process improvement.

Culture and Worldview My father is a cultural anthropologist so I grew up on stories of the strange customs of people from other cultures. A recurring theme was understanding why people do things that seem irrational to anyone outside of their culture. The answers can be found by understanding the worldview and values that drives the culture. As an adult I discovered that the high tech had its own strange culture with a strong worldview applied to everything we did. Those that violated the cultural values were severely punished and some deeds would cause exile or immediate termination.

While readily adopting this worldview, I remained puzzled over many of the inconsistencies that were present. In any culture the rules are seldom recorded, even though they are strictly enforced by everyone in the community. Some communities intentionally document their worldview to help newcomers understand and adapt. The HP Way and the Python Code of Conduct both describe principles that are valued and the expected practices within their communities. These documents serve as a guide and can help newbies avoid cultural blunders.

In order to change a culture you must truly understand what drives it. We need to study the beliefs that drive behavior in order to reverse dysfunction and affect a positive change in high tech corporate culture.

Belief Systems Cultures are driven by a worldview. Individuals aren't aware of their worldview until someone else opposes it. There are key assumptions that are naturally embraced and never questioned. An entire worldview is supported by specific paradigms which each provide a fundamental understanding of the world. A common mental model allows everyone in the community to be aligned because they share the same beliefs. The community believes that all *normal and decent* people think the same way.

We often choose to be part of a community because we share a large number of paradigms with them. We resonate with the ideas that are shared and are repelled by those we don't share. The worldview allows us to readily distinguish between people like us and outsiders.

Let me illustrate this with two examples. The open source community believes that sharing is good. It is a self-evident truth that ideas should be readily exchanged with little thought to collecting money. The

Fortune 500 culture views intellectual property primarily from a security perspective. If sharing information is viewed as thievery it is difficult to embrace any open sharing ideals. These two paradigms are directly in opposition with each other. Which one is right? It will depend on which community you are drawn to. Once you make this choice, it becomes very difficult to function under another paradigm.

Another example is related to our view of competition. Some cultures promote sharing of information and people are rewarded for helping others and furthering the cause of understanding. In other cultures, people are rewarded for knowing more than anyone else. Most academic and corporate environments discourage individuals from helping others because of the perceived loss of personal advantage. Many corporate environments are based on a forced ranking of employees doing the same job. Helping someone else might move you from a 62% rating to a 60%. In this type of environment employees may actively sabotage the success of others in subtle ways, hoping to better their standing.

Dysfunctions of a Team Patrick Lencioni wrote a great book called "The Five Dysfunctions of a Team". He describes how unattended problems can create a downward spiral into dysfunction. The problems build on each other until the team becomes unable to fulfill its mission. Here's his list of team dysfunctions:

1. Absence of Trust
2. Fear of Conflict
3. Lack of Commitment
4. Avoidance of Accountability
5. Inattention to Results

The goal is to build teams that have the characteristics listed in Lencioni's book:

Signs of a Healthy Software Team

1. High level of personal trust.
2. Engage in unfiltered conflict around ideas.
3. Everyone commits to decisions and plans of action.

4. Hold each other accountable for delivering against those plans.
5. Focus on the achievement of collective results.

For deeper understanding of team dynamics, I highly recommend reading "The Five Dysfunctions of a Team".

Role of Architect Now let's build on Patrick Lencioni's ideas to highlight the areas of focus for building healthy teams within the software development context. The slow toxin of dysfunction can be reversed by creating an environment that respects both the task and the people.

Software architects are responsible for leading the technical development of the project. A key aspect of this job is to build teams that are capable of doing the technical work. The architect is positioned within most companies to have a profound influence on the culture of the team. They are trendsetters - they set the tone of the cultural norms that are expected within the team. In a sense, they embody what *normal* looks like.

Your job is to create an environment where each of these healthy team dynamics can build on each other: Trust, Courage, Dedication, Focus, Clarity. This requires a demonstration of true leadership so invest the influence that you already have to make improvements for the benefit of the entire team. This may seem like a secondary goal but healthy team dynamics don't happen without intentional effort and consistently great software only comes from healthy teams.

Attributes of Health

Organizations vary dramatically in performance. The most admired company at this time is probably Google. Their level of productivity is phenomenal and unequaled. This is a direct result of their culture because they can recruit the very best, place them within great teams, and then unleash them on complex problems.

Jim Collins wrote a classic business book entitled "How the Mighty Fall". He described how successful companies have no guarantee of remaining successful. A great company can forget to do the very things that made it great in the first place. When this happens the company will begin a slow and inevitable downward spiral into dysfunction. If left unchecked, a Fortune 100 company may close its doors within a few years.

Although you can't address the needs of a whole company, you can bring light and health to your local team. The Software Architect role requires more than technical skills, it requires leadership. Consider the task of building your culture a meta-project that supports all of your projects. Seek to leverage the culture beyond the scope of a single project.

There are several areas of culture that determine the amount of leverage that your projects will have. Each of these areas should be designed to give you optimal results. Think of each of these areas as a cultural pillar. Together they provide a high-performance platform for your projects.

Trust There's a lot of talk now about trust, and for a very good reason. We live in a low-trust world. Some have pointed to government and corporate scandals as the cause but everyone agrees that trust is a scarce commodity everywhere you look. Stephen M. R. Covey has written two books about trust and begins "Smart Trust" with a fascinating discussion about the direct correlation between trust and prosperity at the national level. The higher level of trust within a society (and the corresponding low level of corruption), the higher the productivity and prosperity. This dynamic can be seen at every level down to states, cities, companies, teams, and families.

A high level of personal trust is a basic requirement for success because without trust everything else is in jeopardy. Lack of trust is the beginning of the dysfunctional spiral for teams so this is also the place to start when building a healthy team. Trust is the foundation of collaboration and an environment where invention is welcomed. Creating an environment of trust will make all other changes possible and neglecting to actively build trust will block every other effort.

Some teams are steeped in fear. Managers don't trust the engineers so they micromanage the smallest details. In return, the engineers don't trust the managers to look out for their best interests. In this environment no one is even thinking about true success anymore - they are simply trying not to lose. There is a significant difference between playing to win and playing not to lose. Confidence ebbs in this type of toxic environment. When no one trusts anyone, people spend a great deal of energy assigning blame to others.

On the other hand, when trust is high, people feel empowered. They want to succeed and they want to make the team succeed. Empowerment restores the desire to make a contribution. It produces a sense of accountability that moves each person to do the right thing. If someone is empowered to make a decision they will work hard to make the right one. Therefore, empowerment is caused by, and is a sign of, high trust. In his book "Trustology", Richard Fagerlin wrote "high trust is the currency of greatness".

Courage to Take Risks As trust grows so does the courage to take risks. Decisions that may take a hundred hours in a low trust environment can be made in less than an hour when everyone trusts each other. Avoiding conflict is a hallmark of low trust. Without open conversations and working through conflicting ideas within the team, plans will be weak and lack commitment from the entire team. Trust creates the courage to act. Without trust there is a lot of energy spent on trying to anticipate every possible problem. Analysis paralysis is a symptom of low levels of trust.

Cooperation can grow when people feel secure. They believe that helping others will benefit everyone. Secure people are friendly and cooperative. Fear causes people to act with suspicion and hostility. When you find a team producing great results you can bet that the people on the team are eager to help each other. Teams that are motivated primarily by competition and self-interest will inevitably descend into dysfunction. It takes courage to step out and propose new solutions and try new things. The courage to innovate only occurs broadly in an environment where individuals feel secure and "safe" to take risks.

Sharing information is a sign of trust. When people are insecure they view holding information as holding power. They are reluctant to share what they know for fear that their power might be diminished. This is disastrous in software development.

It takes a lot of trust for team members to learn from each other. They must truly believe that exposing limited understanding to a co-worker will not hurt their reputation. This can't happen when people are fearful that reaching out for help might create a negative perception of their technical skills.

Taking ownership of problems also requires trust. Empowerment helps people embrace the responsibility to find solutions - this takes courage. Trusting a teammate enough to ask for help can be the boost they need to overcome a difficult problem.

To truly foster innovation requires that people feel safe enough to take risks. In an environment of fear people will focus their energy on protection, making sure they don't lose. This is entirely different than working to succeed. Innovation inevitably involves some amount of risk and failure. If failure is punished your team will be afraid to try new things.

Dedication to the Team Champion soccer player Mia Hamm said, "I am a member of a team, and I rely on the team, I defer to it and sacrifice for it, because the team, not the individual, is the ultimate champion". Dedication to the team is a natural response to trusting your co-workers and believing in your team's purpose. Accountability then increases as each team member commits to the success of the team.

People have an innate desire for mastery and purpose. As these two desires are coupled to the team's goals, the individual productivity is dramatically multiplied. Craftsmanship often gets lost in the rush to market. It is important for each developer to feel that they are growing in their understanding of software. Define roles for each individual on your project that allows them to work on their craft, in the area of their strength, while producing a result for the team. The most important thing a team can do is learn, it is foundational to everything else. Encouraging craftsmanship will help to align everyone with the team vision.

Culture determines the innovative output to a large degree. The most admired companies in high tech are able to innovate faster than everyone else because their environment encourages and supports healthy behaviors and the employees become passionately dedicated to their company's mission. Google grants their development staff 20% of their time to working on what *they* think will most benefit Google. That is a message of empowerment!

Focus on Problem Solving The ability to focus on the highest priority problems is a sure sign of health in a team. During the course of software development there may be thousands of individual problems to solve. The success of the team requires that the bulk of the focus be placed on the technical issues and resolving them quickly.

Some organizations become so distracted by non-technical issues that the product development becomes an afterthought. Bureaucracy and busy work are frequent sources of distraction. Poor interpersonal dynamics can also suck up huge amounts of time. Pay attention to the "pain" in this turmoil and work on team issues before it destroys your project.

A team can maintain focus only if that focus is supported by the rest of the organization. Building a history of reliable product delivery builds a reputation in the larger organization which in turn buys breathing space for the team to focus on building skills and solving problems. When the team isn't allowed to operate with this degree of autonomy the productivity will suffer. Trying to regain trust entails a lot of time spent trying to garner management support through progress meetings and similar events. This distracts from the real work that must be done and stretches out the schedule.

Clarity An empowered and skilled team still needs clear priorities to succeed. Clarity about the highest priorities will allow your team to excel in their work and confusion about what the true goals are will undermine the teamwork in your group.

Leadership matters because leaders set the tone of the environment. They communicate what is most important and work to gain alignment in purpose across the entire organization. Leaders define the vision and objectives for the organization. There is much discussion in management circles about mission statements but they are only the beginning. A leader needs to communicate not only what is valuable but how those values affect daily practices.

Convert principles into practices. How will your principles be translated into a new behavior? If you decide that some documents are a waste of time, then only write documents that are useful. Or if you believe that requirements only can be discovered by customers using the product, then build prototypes earlier. If you believe that complexity is bad, then enforce simplicity standards.

Building the Desired Culture

Is the focus on culture a waste of time? Is it too "warm and fuzzy" to really be of importance? The human dimension in an organization is inseparable and ignored at your own peril. Often it is difficult to persuade the key decision makers that they need to invest in people - unless they believe that their survival is at stake.

Producing high-quality software that meets the current needs is a complex endeavor and it won't happen with a team that is just logging hours. A cohesive team that is highly motivated to make the business successful is absolutely essential.

Essential Goals This kind of focused motivation doesn't come from handing out common rewards and perks in hope of inspiring inventors to produce something of lasting value. The companies that succeed are the ones that inspire their people to reach beyond what they are currently doing. How are you inspiring those you work with? Are there some simple things that you might do to take your team to the next level?

The ability to reach maximum leverage is based, to a large extent, on the culture of your organization. Do people feel exploited and undervalued? If so, you will never get much leverage on any project. An environment that supports engaged employees will have low turnover as they thrive and grow. A punitive organization will drive out the best employees, leaving bureaucrats and low- performing parasites. When the best leave they will take everything they know about the project with them. You will lose access to their domain knowledge, their skill with certain technologies, their understanding of how the product works, their knowledge of the customers and market, their ideas about future product directions, and their insight into weak areas of the code.

Replacing an engineer on a project can cost the equivalent of 2,000 hours of effort, or more, to recruit, train, and bring them up to the same level of productivity. It is hard to imagine a more costly mistake. Avoid waste by working to keep people fully engaged and eager to contribute to the common goals.

If a positive corporate culture is so wonderful why do so few companies invest in their people? It ultimately boils down to a lack of understanding. Leadership can often lose track of the impact, either positive or negative, they have on the people that work for them.

Crafting Your Culture Culture happens automatically wherever there is a group of people. Culture is often based on a shared history of events that results in the creation of rules. "From now on no code will be committed without passing the unit tests. If you break the build you have to be the build master."

Everyone wants a great culture but it's not easy to create. It requires more than lip service or instituting Beer n' Pizza Fridays and installing a foosball table. Crafting a great culture takes months and requires sophisticated understanding of the politics of power. You can start that process by assessing the cultural elements you would like to change most.

Few companies invest the level of resources and focus required to build a positive culture. Some half-hearted attempts to improve the culture can even make things significantly worse. Many companies conduct "voice of the workplace" surveys but if there is no commitment from upper management to take action in response to negative feedback they are better off to just skip the pointless exercise.

One of the most frustrating experiences in modern life is calling customer service and waiting on hold for a long time and listening to the message say over and over that your call is important to them. Over and over again the customer thinks, "if my call was actually important to them there would be someone available immediately to talk to me". Saying we value something without taking action is meaningless and doesn't fool anyone.

Leverage Team-building Expertise The good news is there is plenty of opportunity to increase the effectiveness of your teams by leveraging the knowledge of several decades worth of research on team dynamics. I've already mentioned several great books on the subject. I also recommend the books that came out of the Gallup research, "First, Break All the Rules" and "Now, Discover Your Strengths".

Investing time and energy into building great teams will not only improve the team's effectiveness, it will greatly increase your ability to recruit top talent. Everyone wants to work on a great team but very few people know how to create one.

Ink Your Manifesto Creating a manifesto can be a powerful exercise for a team. It acknowledges that things are not okay and that we have a responsibility to make them better. It can provide courage for the hard work that must be done and it awakens the inner warrior in all of us.

In daily operations, it is the practices that determine our specific actions. These practices must be anchored in a set of principles so that we choose the right practices. But the principles alone are not enough guidance for our behaviors. The best way to handle this is to have both of them described together. The Agile Manifesto (http://www.agilemanifesto.org) is an excellent example of this in action. They describe both the high-level principles and how these translate into specific daily practices.

Gather your team together and talk about the human and technical principles that should be represented in the team's daily actions. Then create your own manifesto to capture the things you value most. It can be as simple as a checklist but your manifesto will define both vision and action.

Writing a manifesto may reveal areas of conflict that must be sorted out with different team members. This will improve your overall team dynamics by getting past the veneer of politeness that many teams use to conceal disagreement.

Once you have a team manifesto, select three concrete actions the team can take to change how things are currently done and move toward the goals. You might consider assigning champions to specific goals. Schedule time for the team to meet quarterly to review progress and make needed changes to the action plan. The regular review of the team manifesto, current state, and goals can become part of a continuous improvement cycle that will also help you integrate new members as they join the team.

As you work through these values and goals, a team identity will emerge. This is an important part of the process. It builds dedication and clarity that energizes everyone on the team. Each team member should be able to verbalize the unique team qualities in a single sentence. "We are the team that _____".

Best Practice #13 - *Build healthy teams by investing resources, creating a team manifesto, and tracking team goals quarterly.*

Problem Building software is dependent on the culture that operates within the company and its teams. Talented individuals will be hampered in their efforts when placed in a dysfunctional environment. A company that has a low-trust culture will be slower to respond to market changes and will create mediocre software products.

This cycle can be stopped and reversed by understanding the basis of a healthy culture. A healthy software team has five essential attributes that can be attained in any organization:

- Trust
- Courage
- Dedication
- Focus
- Clarity

Solution Start with a commitment to improve the culture within your sphere of influence. Gather your team and use an assessment tool to gather real data on the current state of your team. Look for areas of pain and conflict to identify problems that are caused by fear and lack of trust. Write a manifesto with your team that describes the team principles and practices. Create an action plan to move toward the team's functional goals.

Next Steps

- Read "The Five Dysfunctions of a Team" by Patrick Lencioni
- Take the Team Assessment at the back of Lencioni's book.
- Write a team manifesto of principles and practices.
- Choose three actionable team goals and review them quarterly.

Chapter 14 - Learning

"The noblest pleasure is the joy of understanding."

~ *Leonardo da Vinci*

Learning is a Strategic Capability

Many companies act like learning is primarily something you do at the beginning of a career. Their hiring process rigorously screens for specific technology skills and qualifications. The underlying belief is that only the knowledge that a new hire already has is required to be successful. The truth is that what an engineer knows at the moment of hiring is less important than their ability to learn quickly and continuously. Technology moves so fast that the half-life of specific technology skills and knowledge is probably about eighteen months. Any engineer that stops learning is on a trajectory of rapid decline in usefulness and productivity.

A strong base of knowledge is important. However, a habit of learning allows a person to grow into whatever situation that presents itself. How often have you added a new person to a team at the same time a change in the project rendered their specific skill set unnecessary? It is smarter to hire an intelligent learner than a statically knowledgeable one.

There are two primary skills from which all other engineering skills emanate. Analytical thinking and problem solving enables engineers to tackle and solve complex engineering problems. However, the ability learn quickly and effectively fuels all the other skills that are vital to a successful career. What you know now is not nearly as critical to your success as your ability to learn.

Managing Knowledge to Optimize Results So what does learning have to do with software leverage? If you have ever been responsible for completing a project using a team of engineers that weren't adequately prepared for the task then you know the resulting nightmare. Budgets and expectations are set with the assumption of a skilled staff. When key skills are missing, sufficient time must be budgeted to allow for the required learning.

Making blind assumptions about the skill of a team is courting disaster. The skills required for a specific project needs to be correlated with the skills that are possessed by the team. When a gap is discovered, a plan must be created to allow the necessary learning to take place. This is what we mean by skill management - it is simply comparing the need to the asset and figuring out how to close the gap.

In order to leverage software we need to have skilled developers that are proficient with the technology they are using. The depth of knowledge about the technology will prevent blunders when building the solutions. Both breadth and depth of knowledge will prevent technical debt of many types. High leverage of software is only possible if the residual technical debt remains low. Therefore, the technical skill of the team will have a direct correlation to the possible leverage.

New Era with New Challenges Changes due to the internet are fundamental, not incidental. The growth and maturity of the information available online has changed everything. These changes are here to stay and will only accelerate with time. This must change the way that organizations think about the entire process of learning. This chapter will explore the implications of learning and training in an internet age.

Programming problems are solved instantly. A five-minute search on Stack Overflow and Google will answer many technical questions. Insightful articles and tutorials can teach us how to apply a technology to a specific situation. It has never been easier to learn syntax, solve surface level issues, and even troubleshoot certain classes of debugging problems. A favorite approach to debugging is to cut and paste the raw error message into the Google search engine and click "I'm Feeling Lucky".

These type of problems are related to *Programming in the Small*. Small issues require a minimal effort to resolve. The internet has brought us almost instantaneous resolution of small issues. Dealing with tactical issues no longer requires the bulk of our time as engineers. However, software is more than just programming. *Programming in the Large* requires engineering skill that isn't readily available as canned answers on

Stack Overflow. Dealing with issues of a large scope requires a different level of effort.

Unfortunately, many engineers mistake the ability to resolve minor tactical problems with the capability to build a system. The skills and knowledge required to build complex systems are fundamentally different in nature. In the internet age there is an instant gratification ethic that feeds a belief that we can be successful simply by searching and applying canned answers to problems. Now we see that architecture is emerging as the fundamental limiter to quality.

Never before has there been a greater need for software engineers to master the fundamentals of engineering. As tactical learning becomes universally easy, strategic learning is becoming a critical success factor. Anyone can build an application, but will it stand the test of time? These are issues that rest on the quality of the design decisions.

Sustainable Innovation Most organizations claim to value innovation but the truth is that innovation comes at a price. Innovation can only occur through investing in learning because technology moves forward at an incredible pace. New skills are the fuel for innovation. If the staff stops learning, the innovation engine is running on borrowed time and old technology. Old technologies are often still being used to build new products because the belief is that it is *safe*, only because it is what we already know.

Learning is the fuel for innovation. Invest in learning and you are investing in new products. The typical engineer should spend about 20% of their time exploring some new technology. This should be done in the context of the product design rather than as a vague desire to discover. The type of learning that we need most is tactical and focused on acquiring skill for using critical technology. Learning must benefit the team and not just the individual.

Two Bodies of Knowledge There are two types of knowledge that engineers possess and make them so valuable to the organization. The first body of knowledge deals with general principles and practices of engineering. This understanding takes years to build but it also has a long half-life. For the rest of this discussion we will just refer to this as *engineering skill*.

The second type of knowledge is about familiarity with a given technology. The understanding of how to use a specific framework, language, design pattern, programming hack, or database is just a catalog of tricks. This type of knowledge is easy to gain but has an exceedingly short half-life.

When assessing skills and building a team, companies should balance these two different bodies of knowledge. The long-term capabilities of your team are limited primarily by the engineering knowledge of its members. Technology understanding will allow you to produce results in the short term. Companies tend to overvalue either the short-term or long-term skills. This is often tied to the underlying beliefs about the necessary components for success.

Every team needs to produce results immediately *and* well into the future. So seek to build a team that has the proper balance between veteran developers and new graduates. They each bring something different to your team and both are essential.

The 50 Tricks Philosophy

With a sound engineering foundation it is easy to add new technology skills. A competent programmer can train themselves on a new technology in a couple of days and be productive in a month. Gaining knowledge about a new technology is just a matter of learning the top 50 Tricks that are required to be successful at applying the technology. Each trick is a series of steps that are required to implement a small scale tactical solution using the technology. A trick can be fully learned in a day and applied in an hour once the trick has been learned.

Age and experience affect how we see the process of learning. Older developers tend to believe that the engineering knowledge they already have is the key asset. Some old engineers latch onto a technology and choose to ride it for the remainder of their career. They become increasingly specialized in their expertise and committed to it. This may limit their effectiveness and value to the organization, simply because they will resist new tools and process. This lack of growth often limits their role in the project.

On the other hand, young engineers overvalue their familiarity of the newest technologies. Knowing 50 Tricks will allow you to build a small application, but doesn't prepare you to deliver an enterprise-level application that will run for the next ten years. It certainly doesn't give you the skills that you need to leverage legacy code into a new system. This type of knowledge is only built with many years of experience.

Young engineers have grown up in the age of Agile Software and Nike (just do it) notions of speed. So, do we want rapid or sustainable systems? The answer is, yes. Don't undervalue the deep thinkers or the rapid developers. Every organization needs a combination of "Do it right" and "Get 'er done" philosophies. If a team tilts in one direction it will fail.

Focus on Learning, Not Training Training is about pushing information. It focuses on what you give to employees. This covers everything from mandatory corporate training to in-depth technical classes. This type of training is needed but insufficient to make your organization a leader in software development.

Learning is about pulling information. It focuses on what employees do for themselves. Providing a healthy culture that promotes learning can energize engineers to continue learning and developing. This is the kind of culture that will turn your organization into a powerhouse of innovation.

Focus on practical application. Many academic institutions struggle in this area. Much of the information is either theoretical or outdated. Make sure that people apply the knowledge as soon as they learn it. Make the learning have an immediate benefit to the individual as well as to the larger team.

Look for demonstrations of mastery. A skill isn't really useful until you can show it off. Have each person demonstrate the newly mastered skill to other team members or host a seminar to teach the skill to others. Celebrate each success.

Each person should acquire some new skill each month. If a year goes by without any new skills being demonstrated it is possible that learning has stopped altogether. I've been in some organizations that absolutely stifled learning. They viewed learning as a self-indulgent luxury that they were unwilling to waste money on. You get what you pay for and learning can produce a huge ROI if done correctly.

Create a Catalog of Tricks for Each Technology In order to be competent at any given technology you must learn at least fifty tricks. Why not build a catalog right from the start for the top fifty things that you need to do? This provides a great way to structure the learning experience. A trick is a specific way to solve one small problem. There may be many other alternatives, but to start with, we only need one way to get the desired result.

A trick may involve a few lines of code or a shell command to get the desired outcome. A trick is a standard solution to a common problem. It can be easily documented in a single page of text. Use the template of *Problem, Solution, Discussion* to document a trick. Include the code in the Solution section. Describe the known limitations and implementation tips in the Discussion section.

A catalog of tricks is more like a cookbook than a tutorial. A cookbook shows recipes and is very specific. It gets you to an outcome when you don't have a full understanding of what you are doing. A tutorial tries to build your conceptual understanding of the topic. Learning a technology is best done by creating a tactical understanding first and then filling it in with a better conceptual understanding later.

The catalog of tricks is a great way to bring others up to speed quickly. For example, imagine that your team needs to learn Python. One team member could build a "50 Tricks in Python" catalog that everyone else could use to get them started. By selecting the most important things to learn, developers could be productive far before they develop a full understanding. The code snippets and working example code allow them to mimic the solutions provided.

Each Person is Unique and Learns in Different Ways There are now many excellent opportunities for learning. Encourage each employee to find the ones that are most effective for them.

- Tutorials and technology books
- Mentoring and teaching

- Training classes
- Sample projects
- Pair programming and code reviews

Ask each person to produce and review a written plan for the skills they acquired last month and the new ones they wish to acquire. This one act of accountability will produce great results. Most people want to learn, they are simply waiting for permission. This acknowledges that learning is a worthwhile investment. It recognizes the effort that the person is making to develop new skills.

Learning and Promotion Learning is a core competency for engineers. An engineer that can learn is far more valuable than an engineer that already has the knowledge you need now. Tactical knowledge is so easy to obtain that we should recruit people for their engineering skills and then train them in the 50 Tricks they need.

In a healthy team learning will be a natural response to job demands. People want to grow, they just need a healthy environment. If no growth is happening this may be a early warning sign of danger. Is the individual struggling or is the problem larger than that? A history of continual learning should be a requirement for promotion or for giving new opportunities.

Each person has an effect on the team. Look for individuals that are drawing the team into deeper levels of understanding. Where do you see team leadership emerging? Recognize these people for the important role they are playing within the team.

Skill Mapping

How can we effectively manage the skills that are required for our projects? We need a model for matching the required skills to the current set of skills within the team. Next we need a method for actively managing the skill gap so we can acquire essential missing skills.

A realistic assessment of desired skills will include both the general engineering skills and the specific technology skills. Review the overall project needs and ask what skills are critically needed. Identify the skills that have been available on similar projects and ask if your project is in need of those. This may be different than how your team is currently staffed. It may point out a gap that needs to be filled.

Dreyfus Model for Learning New Skills Any given skill will have different levels of proficiency. The Dreyfus Scale gives us a mental model for how we acquire new skills. It has five different phases that each have some characteristic attributes.

Novice

- Doesn't know much about the subject
- May be familiar with high-level concepts and key words
- Shouldn't be trusted to accomplish any goal

Advanced Beginner

- Basic understanding but needs a simple set of rules to follow
- Can't deal with unexpected situations
- Can succeed at well-defined tasks but needs supervision

Competent

- Able to build recipes for others to follow
- Can deliver reliable results and recover from errors
- Treats every situation the same (can't appreciate important differences)

Proficient

- Views the complexity of each nuance
- Often good at teaching others
- Can quickly solve all problems that arise
- Treats every situation as unique

Expert

- Operates largely by intuition
- Success without thinking
- Highly-tuned automatic responses

The distribution of a population is a bell curve that is skewed toward the low end. For assessment purposes we will combine the bottom two and top two categories. This gives us a simple designation of skill level: Beginner, Competent, Master. We will use this scale for assessing the specific skills for our team. Learn more about the Dreyfus scale at https://en.wikipedia.org/wiki/Dreyfus_model_of_skill_acquisition.

Manage a Map of Skills in Your Organization What skills do you already have? Building an expertise map of your organization is a valuable tool. A table can be used to record the name of the skill, level, and individual.

Table 14-2 Example Skill Map

Skill	Person	Level
Angular JS	Hilda	Master
Django	Hilda	Beginner
Post Gres	Hilda	Competent
Angular JS	Hilda	Master
Django	Bill	Beginner
Post Gres	Bill	Beginner
Project Management	Bill	Competent

This type of map gives you a bird's eye view of the skills that exist on your team. Writing this down makes it very concrete. It can become a point of discussion and review. Now you can consider the skills that you need. Set targets to acquire specific skills.

Identify the type of map you would like a year from now. How will you fill the gap? A year is plenty of time to get the skills you are missing. Having a map also gives you the ability to request specific individuals to learn the skills needed in order to best serve the team. Create a plan and review it regularly with each staff member. Measure the progress and acknowledge growth. Whether you are working on a development plan for an individual or a team it is important to celebrate success.

This method also works well for preparing your own personal development plan. Identify the skills that you consider important to success and define the level that you currently operate at. Project out where you would like to be a year from now. Use this to set your personal growth and learning agenda.

Reward New Skills Most people want to learn and it is critical to support this goal. Some organizations discourage learning to their own detriment. Make learning a worthy endeavor by acknowledging the effort it takes and the value it creates for your team.

If someone becomes a master in some skill, recognize them as an expert. This takes very little effort and creates a lot of momentum for other learning activities. At the very least, take a few minutes in a staff meeting to acknowledge the accomplishment. Imagine if everyone on your team had three to four goals they were actively pursuing.

When a student enrolls in college they expect to spend around 8,000 hours of effort before they can begin to recoup the investment. This is not the type of learning we are seeking. Create a continuous path of small learning opportunities with immediate application. Eight hours of effort should result in visible improvements. Think small, quick, and constant!

Building a Skill Map Managing skills within your organization is a fundamental part of your success. Make sure that you have all of the necessary skills for your project represented somewhere on your team. The best way to do this is to be explicit about the level of expertise required.

The Dreyfus model gives us a great representation of the skill levels. People at the lower skill levels will overestimate their capability. This makes self- assessment inherently faulty. It is better to move to ratings that are based on experience. Skill is created by working with a topic over a period of time and objectively measuring productivity.

According to Malcolm Gladwell, in his book Outliers, it takes approximately 10,000 hours, over a decade or more, to truly master a difficult life skill. This is a good yard stick to help us calibrate the *Expert* level in the Dreyfus model. After ten thousand hours of practice, an intelligent and industrious person should be a world-class expert in their field of study.

Becoming *Proficient* in a complex subject can be achieved in about a thousand hours of study. If an engineer spends their time solving problems, a thousand hours will provide opportunity to examine a thousand different problems in the subject area. One hundred hours is enough time for a person to become *Competent* since it exposes a hundred different problems to the learner.

This yields the following algorithm for rating skill.

- Expert - more than 10,000 hours of practice
- Proficient - more than 1,000 hours of practice
- Competent - more than 100 hours of practice
- Beginner - more than 10 hours of practice
- Novice - more than 1 hour of practice

Note that this technique doesn't connect in any way with calendar time. If you work on a project for five years that used Mongo DB but only spent about fifty hours thinking about Mongo then you are still a beginner. The relevant issue is how many hours was your brain processing and practicing on the related topic. This is what determines your true proficiency.

The easiest way to learn how to build a skill map is to build a skill map for your own personal expertise. This is easier than building a map for the team. Once you see how to do it you can extend these ideas to build a skill map for your entire team. Start by creating a list of the top skills that are important in your work right now. You may be an expert at something that doesn't apply to your current work assignment. Only use what is relevant to the job at hand.

Next, create categories for groups of skills. This will help you establish the most important skills and call to mind things that you might overlook. Your list might look something like:

- Technology - tools, frameworks, general solutions
- Languages - all of the programming languages that are in top demand
- Engineering - general-purpose problem solving skills and techniques
- Leadership - skill that help the overall project succeed

Now select the most important five to ten skills in each broad category. These are the skills that you need. Rate the level of experience you currently have for each specific skill. Create a table of each group of skills and where you rate on the experience. Following is a typical skill set for web developer.

Skill Type	Expert	Proficient	Competent
Technology	Django, Git/Github	Web dev, ASP, Angular JS, Docker	PyCharm, AWS, Postgres
Language	Python, C++, Linux	C#, HTML, CSS, JS	Java, Ruby
Engineering	Process, Refactoring Automation, TDD Knowledge mgmt.	Debug, Architecture Patterns, Monitoring	CI, Deployment
Leadership	Planning, Agile Product defin.	Writing, Business Skill mapping, Team lead	Marketing

Table 14-1 Skill Map

Build a Culture of Learning

A lot of learning takes place when motivated individuals teach themselves, but there are some things that you can do to help this process along. Some activities can be done as a natural part of your project work. They act as powerful group-learning and team-building experiences.

Every project requires about 50 tricks in order to get the job done. Think of these tricks as design idioms or structural patterns. You should identify the fifty tricks that are most important to your success. Build a

Catalog of Tricks. Have someone own it and everyone contribute to it. Make sure that everyone can do all the tricks.

The actions that you take have a significant influence on the entire team. As a leader (whether you have a title or not) you can either encourage or discourage learning within your team. Committing yourself to the goals of constantly building your skills and sharing what you know creates a healthy example for others to emulate. Over time this can develop into a stimulating environment where people have a true desire to learn and grow together.

Best Practices for Group Learning Coding standards are important to allow programmers to understand each other's code. Create an open environment where anyone can contribute to each part of the code if needed. Setting up and applying coding style guidelines is an essential part of this process.

Individual performance standards define the level of skill required to play a certain role on the project. What tricks must you master to become a front-end developer? What are the base requirements for a DBA? Some organizations may even define proficiency tests in order to qualify for a role. Have some expectations in place but tread lightly.

Complexity standards are important for the quality of your project. What level of complexity will trigger an automatic refactoring? This is an area that you should give serious thought to since it will fundamentally limit the leverage that is possible. If you are pushing for high leverage you must keep the architecture simple.

Mentoring occurs in many different ways. Sometimes it happens automatically and other times it is orchestrated. Learning events can sometimes be organized but ineffective. It is best to find natural connections between people. Connect those with a desire to learn with others that have a desire to teach. Reward collaboration and mentoring whenever it occurs.

Build Experts One expert is worth five newbies in any given skill. Of course, this varies between skills. A master DBA may be a beginner at front-end development. Use the highest skilled workers you have available within their area of greatest strength.

Experts are cheap for the results they achieve. If you don't have a master in a particular skill, work to grow a master. It is better to spread the mastery between team members. Three masters in three technical areas is far better than having one person that is a master in all three areas.

Beware of an expert without people skills. One caustic personality is enough to ruin an entire team. No matter how brilliant the person is, they are a liability. One bad person can easily consume the productivity of four to five employees.

Look for leadership because a true master leads. Who is influential with everyone on the team? That's the leader, even if they don't have the title. Encourage everyone on the team to learn and share best practices. Leverage comes as a natural result of having high quality designs with high quality experts leading the charge. A project like this will attract additional talent. The work will get done and everyone will enjoy working together. This team will be respected throughout the organization.

Learning Scorecard Learning is a dynamic target so try not to plan too far in advance. This gives you flexibility to respond to changes as they occur. It may be helpful to think about multiple planning horizons: Week, Month, Quarter, Year. What skills are you hoping to build now or at some point in the future?

A review should start with a list of new skills each person has developed. Celebrate the accomplishment and then ask "what's next?" Build on the past with an eye to the the future. Maintain an inventory of new skills needed by the team. Consider the personnel interests of each individual and look for overlap.

Try to track measurable results. This helps you establish ongoing support for favorable corporate attitudes toward training. Let the scorecard show why freedom to learn is an investment worth making. Show the business what they are getting for the money spent.

Cost and Benefit Analysis Learning has a huge cost. A familiar task can be easily completed in one hour. An unfamiliar task requires a day of learning. This accounts for the huge gap between the top and lowest programmer productivity. It is not that the low performer is an idiot, but simply that they require learning the new skill before doing it. The performance gap is entirely situational.

Learning has a huge benefit. Once a skill is acquired it produces a dramatic increase in productivity. But how does this performance aggregate across a project?

A typical development iteration requires applying 50 tricks. A knowledgeable person can complete the job in 50 hours. A newbie will take 50 days (maybe longer). The take-away lesson is this... Expect a range between one week or two months based solely on who you ask to do the job. Remember that this remarkable difference is due the familiarity with the tricks they are required to do. It is no reflection on the personal quality of the individual. Don't ask a painter to do plumbing.

A more typical situation is a project where some of the tricks are known and other tricks must be learned. This project profile looks like the following:

- 50 tricks required
- 40 known tricks = 40 hours
- 10 unknown tricks = 80 hours
- total project time 120 hours

Please note 20% of the project required 2/3 of the effort. Learning has a cost that is disproportionately large. Even though we might know most of what is needed, we spend two-thirds of our time learning. This happens on most projects so it is worth doing the calculations.

Balance Learning and Doing On the first day of a new project, complete the 50 Tricks Exercise. Get everyone on the team into a room. Make a list of the top 50 Tricks you will need for your next delivery cycle. Hopefully this is only a few weeks away.

Now go through the list one item at a time. Select the person with the most expertise with that task. Assign that person a level in regards to that task. Be realistic here since wishful thinking will taint the results. Remember that an *Expert* requires ten-thousand hours of practice. Use the objective scale to assess skill level. Beware of the tendency for wishful thinking. Here is a sample of what the results might look like.

Database integration - Who is our best expert at CRUD?

- Bill - Always wanted to learn SQL; spent a few hours
- Item 42 - Bill - Beginner - 10 hours

Login screen - Who is our best person at login?

- Cindy - Created 327 login views at last count
- Item 43 - Cindy - Proficient - 1 hour

Proceed through each item until you have an accurate picture of the skills you are missing. Then estimate the learning cost of filling this gap. Maybe it is cheaper to borrow an expert from the outside than it is to grow your own competency within your team. Is this a one-time thing or something you will need repeatedly?

Switching Technologies Switching technologies is hugely expensive, because you must pay for every engineer to climb a learning curve. Unless you are ready to pay a 10x penalty in productivity you should stay with the current technology. Switching technologies must be an enterprise-level strategic decision.

Learning a new technology is a straight-forward task but it requires time. Each member of a team must relearn how to do tricks that they already knew well in an old technology. The old tricks are worth little once we switch to a new language or tool. This learning requires time and you must budget for it. You must budget time for filling the skill gap. This applies to each trick that is not currently familiar to the developer that must perform the trick.

When you do decide to switch, you should set up a team to learn all of the required tricks. It should be their responsibility to teach others what is required in order to be successful with the new platform. In this way you are hoping to use a team to leverage the learning and benefit the larger team.

Be careful about projecting performance history for a different technology. Manager often delude themselves by making assumptions. Most managers instinctively believe that learning is not that big of a deal - but it is. You may want to set expectations for the recommended investment of time. Here is an example:.

- 50% on solving short-term problems on project
- 30% on solving long-term problems and applying new skills on the project
- 20% building skills not applied to the current project

Your future really does depend on learning. Play the long game to build the future you want. Use a regular review to create a feedback loop. Learning is either encouraged or thwarted by policies. You get what you reward so invest selectively for the highest ROI. Provide key resources that people need (time, money, freedom, guidance) and they will learn.

Best Practice #14 - *Make learning a top priority by measuring it, planning for it, investing in it, and rewarding learning when it happens.*

Problem Many companies undervalue learning. The emphasis is placed on achieving immediate projects goals. This forces engineers to learn on their own time. While this strategy may appear to work in the short-term it can be disastrous in the long run. A significant amount of time needs to be invested in learning for a development team to remain competitive.

Many projects lack the fundamental skills that are needed for successful completion. Teams are often unaware of missing technical and leadership skills that are critical to the project. A few people on the team may have many of the skills and other team members may be underutilized because they lack necessary skills. A project can't be run effectively unless the skills on a team are managed properly.

Each team member needs to have a clear picture of their skills and how they can contribute to the team effort. A map is a great tool for managing the necessary overall project skills and how each individual contributes to the team.

Solution Use the Dreyfus model to score proficiency levels for each skill. Use the number of hours as a true indicator of skill level. Build a map for each individual, the team, and the project needs. Look for gaps where new skills must be learned and build a plan for acquiring the necessary skills. Assess who is capable of the highest skill level for the most critical skills.

Skill maps can also be used to manage the ongoing development plan for each team member. Encourage engineers to invest time in developing new skills and celebrate the success. Have engineers train each other to accelerate the learning experience and have people from outside your organization help you learn faster.

Next Steps

- Create a personal skill map.
- Create a skill map for your current project.
- Identify the top 50 Technology Tricks that you need.
- Create a skill map for your team and identify gaps.
- Build a simple plan for skill acquisition.

Chapter 15 - Planning for Leverage

"Waiting for perfect is never as smart as making progress."

~ *Seth Godin*

Flexible Planning

At the start of a software project there is tremendous risk because so little is known. This is the point in the project where uncertainty is the highest. Everything is unclear - goals, customer needs, technical solutions, and tools. This uncertainty undermines our ability to plan and budget.

At this early stage important decisions can have profound consequences on the success of the overall project. A single mistake can sink the project. The primary goal is working to eliminate as much risk as possible as quickly as possible. We must perform experiments that yield fundamental understanding that can translate into a stable project plan. Planning without understanding is a path to disaster.

Learning occurs throughout the course of a project. Essential information will always be missing at the inception of a new project. We can guess and make assumptions but we need experiments to validate these assumptions. Erik Gries, in Lean Startup, gives a model for how a startup company can experiment to learn more about the business it is creating. This also applies to Fortune 500 companies that need to learn about the new business opportunities being pursued.

Leverage is a fundamental part of project planning. Leaving leverage as an afterthought will drastically reduce the amount of reuse. Only by explicitly pursuing opportunities to leverage will you be able to maximize the development productivity. All of this requires a planning method that is very different than the standard process.

Big Bang Planning   Traditional software projects begin by documenting the project requirements. Based on these requirements a rigorous project plan is created for a single large delivery that is either months or even years away. Let's call this planning method Big Bang Planning, since it is built around one large payload built over an extended period of time. We will be advocating a different style of planning built around the principles of Agile Software Development.

The fatal flaw of big bang planning is that it assumes that the world is a static place and that we have complete knowledge of the project from the start. Both of those assumptions are false. It is impossible to comprehend every element of what the project will entail in part due to an incomplete understanding, but also because important things will happen throughout the course of the project.

With big bang planning the changes and additional learning are largely ignored and produce a great deal of wasted effort. There is no good way to make adjustments during the project because changes are viewed as compromising existing commitments. It is important to understand that these commitments were made prematurely and based on flawed understanding.

Companies that rely on big bang planning typically create an adversarial relationship between the development staff and the business managers. The business people ask for the impossible and the dev team fails to deliver. After a few rounds of this dynamic there is a complete loss of credibility and trust.

Rigid thinking during project planning will always create this dynamic. Key decisions about customers, technology, product features, and implementation details need to remain fluid for as long as possible. In programming terms, we would call this *late binding*. It lets us lock down certain details while letting us keep the available options open. The more you can do this the more successful you will be.

Flexibility Creates Responsiveness How do we move forward with necessary planning and still remain flexible? Focus on short-term clarity that enables design decisions. Some assumptions won't yet be validated and these will block vital decisions. Create simple experiments that allow you to answer these questions and give you real data for making the key design decisions.

Some of the things you learn will surprise you. Many of these surprises will require rethinking your assumed development plan. Be willing to make adjustments to realign the long-term direction. Don't fail to believe what you just learned; you will be tempted to reject the findings but they are a clear warning that your assumptions might lead you down the wrong path.

Monitor all of the new learning that is taking place and changes that are occurring around you. Frequently this new information will demand a response. Seek to incorporate the new learning into your existing plans. By adjusting your plans in small ways as you learn, you can avoid big adjustments later and keep from losing work.

New Learning There are several areas where you can expect to have a large amount of learning.

- Customer needs - It takes a great deal of time to fully appreciate the needs of your customer. Expect to spend 25% of your project time when entering a new product domain.
- Solution requirements - The customer needs must be translated into specific product features. This must be validated by allowing customers to interact with early versions of the product.
- Technology tricks - Create simple prototypes to prove out technical tricks independent of the product code. Doing this early makes the rest of the dev cycle predictable.
- Implementation hacks - Many specific chunks of code are reusable. Begin building an inventory of code snippets to draw from during the construction phase.

Follow Opportunities There are many different ways to recycle understanding in a software project. Be on the lookout for how to take advantage of these opportunities. You may be able to leverage an entire framework by refactoring existing applications. It may be possible to evolve a particular design to address needs that are different than its original setting.

Existing redundancies can indicate opportunities for leverage. Designs or code components that are similar to one another should be combined into a more general purpose design. Leverage the work that was previously done to build a new design that is stronger and simpler. Now this improved design can be leveraged in the future. Never miss the chance to simplify something that looks complex.

New solutions may come from sources outside of your team. By remaining flexible in your planning you will be ready to embrace new components that may solve problems without having to design them from scratch.

Present the team with a list of key problems that need to be solved. Allow the opportunity for engineers to mull over the issues. Some solutions will emerge naturally through subconscious processing rather than a direct project plan. This only happens if you leave the door open for the possibility.

Functional Breakdown

Planning that isn't tied directly to the final solution tends to become focused on the business needs and avoids the feasibility of the actual development. It's important to merge the business needs and the solution architecture, so planning should be built upon the architectural design. This emphasizes the development of the real system, instead of only what the company states it needs. Both of these pictures are important - but must be kept separate. The planners must work to close any gap between what the development team can produce and what the company needs.

A block diagram is a good starting point for estimating the development effort. The architecture can be used as an inventory of the required development tasks prior to release of the product. An estimate can be prepared for each subsystem and then the overall effort is totaled.

You can figure out what development assets are available from previous projects and use this information to figure out how to apply these assets to the new design. The leverage is a result of avoiding work by applying previous solutions to new problems.

System Decomposition Every system can be decomposed into three to five subsystems. This represents the top level architecture of the system. If there are too many subsystems it may mean that the system should be re-leveled to create fewer top level components.

Another level of architecture can be created by breaking down each of the top level components to reveal the details. Starting with a good architecture that covers two levels of detail makes it easy to make detailed estimates of the development effort.

A simple and clean model of the architecture also reinforces the overall design for each of the engineers. Unnecessary details are obscured to focus on the large interactions within the system. A great design should be able to be drawn on a napkin by any member of the development team.

A clean top-level design will also help you identify critical interfaces. Everyone can see the overall data flow within the system and get a quick idea of where the different data elements live. A system has a natural tendency to to disintegrate over time. This results from each part of the system being developed independently. You must maintain the system integration throughout development to avoid a costly "big bang" integration.

Building Blocks for Leverage The block diagram composed of approximately twenty components provides the backbone for our estimation. Each component represents some portion of the final system that will be delivered. This will require some amount of work to be done.

Some of this work is similar to what was done on previous generations of products. Other work will need to be done from scratch. The similarities with other products gives us a chance to recycle our understanding to build the new solution. Identify the opportunities for the level of reuse for each of the components.

Interfaces regulate component interactions. Well-defined interfaces amplify the level of reuse that is possible. Strong interfaces make it easy to encapsulate the data for the components, which lengthens the useful lifespan of the component. Creating new solutions that use these components is often the highest form of leverage that is possible.

Each component has a life cycle. It will need to be defined, designed, implemented, and tested. Once again, you can look at each subsystem and see which of these tasks is mostly complete and which ones need a large amount of work.

Quantify Desired Leverage The effort estimation can be done one subsystem at a time. For each of the twenty building blocks you will decide on the assets to leverage and how much work is required to finish that subsystem. This should include all of the life cycle aspects required to bring it to completion.

Create an estimate for remaining work to be built on the existing assets. There may only be minor changes that are required to make it ready for the new system. Or, in other cases, it may require implementing a component completely from scratch.

Build all of this information into a map. Create a table that captures the define, design, code, and test effort for each component. Estimate the hours of development effort needed for each part of the work. This is the beginning estimate for your project.

Levels of Leverage Leverage is not a black and white decision that happens at one time. There are many different assets that might be reused in various ways. The question is never, "Should we leverage?", but rather, "How will we leverage?". Leverage can planned on a module-by-module basis. A complete redesign of a component still allows a small degree of leverage.

Understand the different levels so that you can estimate the amount of reuse that you expect to achieve. For this planning exercise, we will equate amount of leverage with the cost saving to the budget. For example a component that cost 100 hours to build with a 30% leverage should cost 70 hours.

Leverage Type	Description	Savings
engineering leverage	Use new tech to solve new problem	10%
problem leverage	Use new tech to solve old problems	30%
design leverage	Use existing designs to rewrite code	50%
code leverage	Use existing code with modification	70%
code reuse	Use existing code in a new context	80%
maintenance leverage	Fix bugs and enhance	90%

Tracking Your Progress

Now we are ready to begin tracking the actual progress during development. Start with the initial estimate that is based on the architecture and the leverage you expect to achieve. Throughout the project we will be updating the actual progress made and updating the estimates for the remaining work.

People naturally make a connection between the effort that is remaining and the calendar time remaining. This assumes a linear progression that is predictable and can lead to serious misunderstandings if a gap is allowed to grow between actual performance and estimated performance.

The calendar time remaining is based on the velocity of the work that the team is able to achieve. If the team is consistently delivering less than the projections indicate, you should enlarge the estimates for all of the remaining work. Maintain consistency between effort and calendar to prevent misunderstandings.

Create a spreadsheet to track progress. For each component record the actual effort that has already been invested and the remaining effort. Update the remaining effort to reflect all of your current knowledge. Now you can sum both the total effort invested and the remaining effort. This gives a very reliable indication of when the product will actually ship.

Reconcile the current plan with reality weekly. Any change in the schedule should be the result of actual learning that has occurred. This learning will typically be reflected in small changes throughout the course of the project. Large changes should only happen when something dramatic has been learned.

Progress by Implementation Level Building a new system typically proceeds in layers of development. The first step is to get all of the core functionality working. Then the different types of errors are handled. It is much easier to get the happy path working first before considering all of the errors that may occur. Trying to address errors too early in the design of each component can result in a more complex design.

I would estimate that about half of the implementation effort goes into addressing issues related to error handling. So this gives you a way to track your progress through each feature. When you have implemented a solid handling of the desired functionality then you are really about half done with the finished feature. You can use this to estimate the remaining work throughout your project. This is another area that you will want to calibrate to match your own development practices.

Progress by Component Update the progress of each component on the block diagram weekly. Record the effort invested and project the trajectory based on updated knowledge. Did the job just get smaller due to a new simplification or did it grow due to a new requirement that was discovered?

Focus on remaining work rather than who is to blame for not seeing the issues earlier. Hold your current estimate loosely. Remember that each week you will learn something that will alter the estimated release date. The progress spreadsheet should show a count down to release. The remaining workload should be dispatched in a linear fashion. Any blips should be due to new knowledge that is acquired, not variations in productivity.

Adjust estimates weekly to keep the changes small. This will make the project remarkably predictable and will build credibility and trust throughout the organization. It allows you to demonstrate that you deliver on your commitments.

Calculate Work Remaining The entire estimation tool only works if you give it good data. Putting in initial predictions that are too optimistic, or refusing to change the estimates will ruin the usefulness of the results. Honesty is required to keep the actual progress aligned with expectations.

If you are in an organization that doesn't tolerate changes in estimates then don't ever publish the component estimates. Conduct business conversations based on estimates that are about double of the bottoms-up estimations. This outer estimate should be achievable over the course of the project.

Give the progress high visibility within the development team. Promoting awareness of how each person's contribution affects the overall effort will foster a degree of accountability and increases the desire to win.

Update the progress tracker with new info. Review and revise each of the twenty primary building blocks weekly. Most of the time very few components will change. Only those where active development is occurring should see significant change. This review will often take less than an hour and can be a great part of a team meeting.

Measure Velocity Weekly Velocity is the amount of work that is done by the team over a week. This should match the development team's actual effort. A five-person team can do about 150 hours in a week but not 600 hours. The estimates for future hours serve as place holders until real measurements replace them. For example, we thought it would take 200 hours to build a component that ended up requiring 150. The amount that should be recorded as effort is 150, not 200.

Time remaining should be adjusted based on the actual rate at which things are done. If we consistently run two times too optimistic (or pessimistic) then this should be factored into all future estimates. It does no good to try and tilt the scale. The truth is best digested in small doses.

It is critical that the time scale used for future estimates is consistent with the scale used to track actual investment. Whether you use only engineering time (removing meetings, etc.) or total time spent by the team, it doesn't matter as long as you make the past match the future. This also applies to the units used to estimate tasks. Milli-fortnights (for example) will work as units of time if you use it consistently throughout your plan.

Change estimates that are clearly wrong, no matter how fond you are of the first estimate. A choice to add or remove features can have a large impact on the work remaining and these changes should be factored in as quickly as possible. Then everyone can see the impact that the change will have on the project immediately. The goal is to maintain a consistent view of the estimated completion for the current plan. As the plan changes it is necessary to update the information and project the new end date.

Controlling Scope

Most groups consider estimation to be important in creating predictability within the project. It is good to be able to measure when a project will be done but it is far better to control when it will be done. Project managers often expect that they must have a certain amount of functionality by a certain date so the discussions center around whether the dev team will be able to do the job.

This sacrifices an important degree of freedom in the planning. The scope of work to be done can, and must, be controlled. There is no one correct answer for functionality. Constant evaluation of how to get the largest ROI for the project is essential. Controlling scope is the most important leverage that you have in planning. If you must have certain features sooner then figure out the scope that will give you what you want.

Business planning is separate from technical planning. The project planning process can often degenerate into adversarial interactions between the business planners and the technical planners. This is very counterproductive and harmful to the project. It creates a win/lose scenario that blocks the kinds of trade- offs that are essential for the success of the development.

Tracking Over Estimation The most important step in improving the overall planning process is to emphasize actual progress made over future work that must be done. Use the past history to validate the future plan or change it. Estimation assumes a fixed deliverable - this is almost never the case. Instead, consider alternative goals that can be completed rapidly.

Evaluating other goals will reveal the true priorities of the organization. This will promote healthy discussions that engage all of the parties in useful collaboration. It will close the expectation gap by giving everyone the same picture about the choices that must be made. In essence, a software project is simply a massive exercise in problem solving. More people contributing to the solutions will improve the ultimate outcome.

Recording progress weekly and factoring in new learning will revolutionize the way your team operates. If developers know that any work completed by Friday afternoon will be viewed by the entire organization then they will push for closure. This is only natural and doesn't take any extra effort. Just make the goals for success clear and people will set them as personal targets.

Embrace Reality Wishful thinking and unfounded optimism can't persist in a data-driven environment. People are forced to accept *what is* rather than *the way they wish it to be*. I have seen this transformation occur numerous times throughout my career. Teams steeped in years of dysfunctional interaction can be turned around in a few months as trust is built throughout the organization.

This could happen in your organization. You don't need an official mandate or title to begin tracking progress. Once you begin presenting graphs showing real progress that drives real planning it will attract attention. No manager could ignore the informative gathering of data that brings clarity to the project.

Each week you should maintain the list of work that is fully complete and update the time required to do the work. Then update the list of tasks still remaining prior to releasing the product. Find and add all new tasks to the list. Estimate the amount of effort required to complete the remaining tasks based on the team's past history.

Understand the impact of changes that are occurring within your project. Are you happy with the changes that occurred this week? Is everyone aware of the impact of decisions related to scope? If anyone would be shocked or disappointed there is room for improvement in the planning process. Calculate the new release date based on everything that you know now.

Control the Scope Having clear product goals and finish lines is the most important aspect of project planning. Without them the business and technical people end up making trade-offs that are inconsistent. Lack of clarity results in confusion and conflict throughout the organization. Clear goals will guide everyone as you make adjustments to the product scope. You can easily reject or postpone features that aren't required in order to complete the highest priority goals of the project.

You must also believe the progress tracking data and corresponding estimations. If you build a tracking system but don't believe what it is telling you, the opportunity to take timely actions is lost. Having data is only valuable when you are willing to act on it.

Evaluate the previous changes that were made to the project plan. Did they have the result that you expected? If not, why not? How can you make changes that are more effective?

Use Learning to Update Estimated Completion Everything you do on a project deepens your understanding. Integrate this learning back into the project plans. Each week you should become better equipped to overcome the challenges. As the project proceeds the surprises should be smaller. Construct experiments to answer questions of technical and business feasibility as early as possible. Don't save big risks for late in the project.

By halfway through the project your scheduled release should stabilize. Large surprises indicate that you didn't address a key risk early enough. Learn from this experience by making decisions about how you will handle this in the future.

Build new models of leverage as the understanding grows. Find ways to get more done with less effort. Perhaps you can create a framework that will reduce your effort on the next project or merge two products into one? Refine the planning process with each development cycle. Engage the engineers and business partners in a collaboration about how to deliver more functionality to the market more rapidly.

Best Practice #15 - *Adjust plans throughout development to capitalize on new learning and track progress weekly.*

Problem Planning software projects is frequently ineffective when it comes to managing the unknown risks. Most projects start with planning that is fundamentally flawed by making early commitments that are rigidly followed throughout the course of the project. This can have a catastrophic impact on the project because problems will only be revealed at the end of the project when effective action is the most expensive.

The astounding failure rate in the software industry indicates that most companies follow a similar course regarding project planning. We need a model of planning that has the following attributes:

- Reliable prediction of release date
- Highlights key technical risks so that they can be addressed early
- Robust tracking throughout the course of the project
- Lightweight to oversee and use
- Encourages collaboration between business and engineering staff
- Allow control of scope during the project
- Highly visible progress
- Progress is based on actual product architecture

Solution Create a functional breakdown of the product architecture. Estimate the effort that is required to complete each of the twenty most important components and other critical engineering tasks. Track the progress of completed work weekly. Record actual effort of work done and estimate the work remaining. Then post the results to align the work and demonstrate progress.

Each week a release date is calculated from the progress. An analysis of what new things have been learned is applied to the project plan and any required adjustments are made to meet the overall business goals. Small adjustments are made to the project plan on a weekly basis to prevent big surprises at the end and to integrate learning.

Next Steps

- Create a high-level system partitioning with two levels of detail.
- For each of the top twenty components, decide on which assets can be leveraged.
- Estimate the engineering effort required to complete each component.
- Build a spreadsheet for tracking progress with a table tracking "component", "done", and "remaining".
- Update information weekly.
- Decide on changes needed to meet business goals.
- Keep a log of what was learned about planning the project.

Appendix A - Build Your Own Complexity Measurement Tool

Enumerate Source Code

A system is built from source code. This is any code that a human must write and maintain. A tool might be written in 100 lines of code to emit an XML file of one million lines or it could be an algorithm written in 100 lines of C#. Both of these examples only account for 100 lines of source code, since the million lines is automatically generated and therefore not source code.

Complexity should be measured to reflect the mental burden associated with the maintenance of the code. The first dynamic that causes complexity is the sheer size of the source code. A project of 100,000 lines is more than twice as complex as a project of 50,000 because complexity grows exponentially with size.

Almost all source code can be organized as files with lines of text. By counting the files in our project, the lines in each file, and the total lines of source code in our project, we can generate a basic complexity measure.

For a really cheap measurement, create the following shell script to count C++ code.

```
find . -name *.cpp - *.h | wc -l
```

This script will give you an overview of your source code and it is certain to reveal hot spots that should be addressed. It is based on one key assumption, that files contain logic and a very long file is an indicator of complexity. This may be the most sophisticated measure you will ever need.

If you need more metrics, you can build a tool that is based on a deeper understanding of code complexity. The first task is to find all of our source code. Here's a tool that lists all the Python files in a directory and reads the text from each file and removes blank lines.

```
def python_source_code():
```

```
    return glob(d+'/*.py')


def read_source(filename):
    text = open(filename).readlines()
    lines = [x for x in text if x.replace(' ','')]
    return lines


def list_code():
    for f in python_source_code():
        for line in read_source(f):
            print(line)

list_code()
```

Measure Module Size

You can easily adapt this code to your environment and language choices. Building your own tool can be far better than using a pre-built tool because it reflects your own specialized and dynamic needs. The next step in building a complexity tool is to measure the size of each of the files in the source code.

```
def module_size():
    for f in python_source_code():
        source = read_source(f)
        print(f + ' lines:' + source)
```

Calculate Complexity

Now we can account for the exponential impact of size on complexity by converting the print function to a complexity term. An exponent of 1.2 is used to account for the interactions of each line with the other lines in the file. This is a parameter that you can tune to reflect your situation. The value could be set from 1.0 (reflecting no internal interaction) all the way to 2.0 (reflecting that every line affects every other line). A good starting point is 1.2 until you have more data.

```
def module_complexity():
    for f in python_source_code():
        source = read_source(f)
        complexity = len(source) ** 1.2
        print(f + 'complexity:' + complexity)
```

The next development is to look at the functions within the module. We will identify the length of each function within the source code. The is_function code locates the functions within the source and the list_functions logic calculates the lines within the function.

```
def is_function(line):
    if line.strip().startswith('def'):
        pat = compile(r"\s*def (.*)\s*\(.*")
        name = pat.sub(r'\1',line)
        return name


def list_functions(lines):
    start = 0
    for i,line in enumerate(lines):
        function_name = is_function(line)
        if function_name:
            name = function_name
            print(name, i-start)
            start = i
    print(name, i-start)
```

Estimate Non-linear Complexity

The next step is to assess the relative cost of each of the functions. This time we use a 1.3 exponent to reflect the deeper interconnection of the lines of code within the function. This setting could reasonably take on any value in the range of 1.1 to 2.0.

```
def function_cost(name,size):
    cost = size ** 1.3
    print('    %-26s %8d %8d\n' % (name, size, cost))
```

Now that you have seen the individual pieces we will integrate the whole thing together into a program. Our tool will do the calculations and build the total complexity measurement in memory, then print the summary.

```
def count_functions(lines):
    results = []
    start = 0
    for i,line in enumerate(lines):
        function_name = is_function(line)
        if function_name:
            name = function_name
            results.append(name, i-start)
            start = i
    results.append(name, i-start)


def list_modules():
    return [ f,read_source(f) for f in python_source_code() ]
```

Note, count_functions measures the size of each function and returns the data as a Python list. The function list_modules returns a list of all the source code text. The calculate_complexity function now goes through this list data and converts it into a complexity metric.

Estimate Module Complexity

Now let's look at calculating the cost of each module within our system. The cost of each function within the module is summed. Then an exponential factor is applied to the total. This compensates for the extra penalty of many functions within the module.

```
def module_cost(lines):
    module_cost = 0
    summary = '\n'
    name = 'module'
    for x in count_functions(lines):
        cost,details = function_cost(name, x[0])
```

```
            summary += details
            module_cost += cost
            name = x[1]
    return module_cost,summary


def cost_of_modularity(lines):
    size = len(lines)
    return (size/2) ** 1.1


def complexity(filename, lines):
    num_lines = len(lines)
    cost, summary = module_cost(lines)
    cost += cost_of_modularity(lines)
    s='%-30s %8d %8d %s'%(filename, num_lines, cost, summary)
    print(s)
    return (num_lines, cost, summary)


def show_complexity():
    print('File                          Lines  Complexity\n')
    total_cost = 0
    total_lines = 0
    for filename,source in list_modules():
        num_lines, cost, summary = complexity(filename, source)
        total_lines += num_lines
        total_cost += cost
    print('\n%-30s %8d %8d' % ('total',total_lines,total_cost))
```

You now have your first complexity metric and can go all the way from source code to a summary complexity map of your code. You can control every assumption and code in additional attributes, as needed. Analyze your code and correlate your knowledge of the coding effort with the complexity metric from the tool. As you gain insight, make adjustments to the tool to reflect a deeper understanding. For example, many imports add system complexity so build an import detector and a penalty term for each import.

Appendix B - Testing Automation Interfaces

Build around testing scenarios

Each feature in your product should have a test case. Each line of code should represent some requirement in your software. A test case can be implemented in a single line of code. This give us the following guideline.

```
Lines of product code = Line of test code
```

In other words, if you count the source code in your project you should have about 50% test code. This seems like an unreasonable and radical requirement for those who do not test there code. But this practice is well supported by many studies throughout the industry.

Do not code complex test. Instead, focus on simple assertions that can be easily written as a single line. Pick isolated things that you require to be true. Hard-code everything and pick as many test cases as you need. Set time goals for writing tests. This forces you to do it, but keeps it simple. My personal goal is write a test in 30 seconds.

Anything that is a function should be called from a test case. Why would you wish to have code in your product that is not tested? For more interesting functions you need to have multiple test cases. But even a single invocation will detect a lot of catastrophic failures.

I have been doing unit testing for twenty years. I find that the simpler tests are better. They break less, and require less effort to fix. I have seen countless team abandon unit testing simply because they would only pursue a glorious version that tested every possible path.

This is a serious mistake that could undermine your projects success. My advice is simple. Build a simple test case for every line that needs to work. The lines that you do not care about should be removed from the source code.

Software is build by compositing low level features into medium sized ones, and finally into top-level programs. When tests are written with the product code they should start at the lowest level feature. As you turn on a small scale feature you code a test to verify what you intend.

This causes the development to go from the inside (lowest level) to the outside (application level) of the software. A system built in this way will generate thousands of test cases that are a byproduct of the development itself. You try a simple thing and save the code for later execution.

The entire battery of thousands of test cases will ensure that the last change you made did not cause any new problems. Every hour you can run your tests. Whenever anyone commits code you can run your tests.

This kind of testing is essential to refactoring. If you do not develop unit test, then you will never have the confidence to refactor your code. If you cannot regularly refactor, then your software will decay into disuse within two years. Your choice!

Stimulate the interface

Every subsystem in a piece of software does something interesting. Each component serves a purpose. The strongest abstractions in the system are verb/noun pairs.

The nouns in the system are the types of data objects that are being saved, recalled, and acted upon. They are typically represented as models in the program and persisted as database rows.

The verbs in the system represent the operations that are performed. They are typically implemented as functions or method calls on objects.

All of the business logic in an application comes down to these interactions between nouns and verbs. Examples would include:

- list customer add order build inventory import doomsday_device launch
- missiles signal surrender

Most applications are built around one to four key interfaces. These are the logic points of interaction with your application. An automation interface allows you do have an important conversation at that point with your application.

Consider the last app you worked on. What was the most important interface in that app? Now identify the nouns and verbs you worked with.

Would not it be cool to have a language that would let you converse with your app right there?

Imagine being able to capture the traffic at that spot and save it to a log file. You could have a transaction log that gives you a quick playback of the history as it existed at another point in time. Capture and playback gives you amazing control over running specific scenarios within your system. You can also capture sequences and move them to a different system.

An automation interface gives you a wide range of options. Systems that have explicit automation interfaces exposed are far more flexible than other system that cannot be controlled from the outside.

Exposing an interface and implementing a text language to drive it is extremely simple. I recommend that everyone consider building at least one key interface to their application. It will come in useful in ways that you cannot even anticipate. It is not necessary to predict all the ways that you will make use of the interface.

A domain specific language (or DSL) is a term that makes all this sound sophisticated. But the practice of automated interfaces is really just common sense. These command interpreters can be built in any language in a day by a competent programmer.

Testing with Live Data

Most systems deal with some type of data. These objects may to Orders, Students, Launch codes, Sensor readings. No matter what the data types represent, there are certain to be some core operations that you will need to do. CRUD (create, read, update, delete) is one set of operations that you will probably need. Your application may also call for other operations (eg. execute, recover, rollback, commit).

Build a command language to capture these types of operations on your data. Implement CRUD as a starting point but extend your language to cover the specialized verbs that you need most. This language is your injection point for trying out your test cases.

Use your DSL (domain-specific language) for making requests of your system. Remember that a test case is just a request/response pair. You

DSL should allow you to implement a single test case on a single line of code.

At this point you are building a conversation with your system. It is important to design safety switches, that prevent altering the system behavior within your production environment. This can follow two main patterns: sentinels and fake data. The sentinel pattern involves preventing certain operations from occurring on certain systems. For example, don't send notification emails from the staging server, or allow actual orders to be sent only from the production server.

The fake data pattern marks all of the fake data inserted into the system. This lets real data and fake data to mingle within a system without being confused. Certain commands can be prohibited on the fake data (or the real data). For example, "never send email to a fake user", or "do not delete real users" would fall into this camp.

Once you have a command language that interacts with your objects, you can start telling stories. These stories could be composed of transactions that alter the state of the system. For example:

- add character Evil Queen
- add character Cinderella
- build Castle
- move Evil Queen to Castle
- show Castle
- list characters

Executing transactions advances the state of our stories. Other transactions can reveal the current state. Imagine the possibilities for build testing scenarios. The individual test cases each combine a request with a response, but they also change the state of the system. Some of the test cases just verify that the system got to the correct state at the right time.

Appendix C - Component Encapsulation

Encapsulation and data hiding

The structure of a component determines how versatile it is. If a component is flexible it will last for a long time. Rigidity will limit the lifespan of the software because any change will break it.

We need to be careful not to overbuild the system by making too large of an investment up front. Our goal is to do just enough engineering to meet the need at hand. Do this by stopping short of building any functionality or flexibility that is not needed immediately.

There are several areas where we need to apply just enough engineering. When building a component we want to only expose what is needed to other parts of the system - everything else should be hidden from the outside world. Exposing additional data and operations is easily done later, but the opposite requires a lot more work. Hiding data and operations that were exposed is a troublesome task which is likely to break code.

The design of a component needs to match the complexity of the task. It should be just sophisticated enough to meet the current challenge. When new challenges arise as a result of new features, then the current design can be extended to tackle the new requirements. Never build a design that you are not ready to use the same afternoon.

Try to envision the simplest possible version of each feature. Then simplify it some more. Don't build complex design rules and logic. Instead, rethink the actual requirements and change them only if they produce a more elegant solution to the problem. Be sure and communicate with the key stakeholders to help them understand the true cost savings that can be had by making the requirements less severe.

Every interesting problem in computer science is solved by adding a level of indirection. However, by waiting until you have a pressing need before adding any indirection you avoid wasting effort. Many engineers are quick to put in a general purpose solution when it's not truly needed. For example, building a plug-in architecture when you only have a couple of things that will use it is a waste. Instead, wait until you have at least four different consumers of an interface before you build the mechanism.

Generalizing a solution too early will cost a lot more. Every solution has design parameters that are impossible to anticipate. Wait for at least three instances of a particular design element before you try to abstract the commonality. Build flexibility just in time to use it.

Extending functionality

There are two competing desires when it comes to extending the code that you write. We want to complete the work that we produce to eliminate loose ends. We also want to allow easy changes for later. It is possible to support both of these goals simultaneously.

We need clear guidelines about how to extend the functionality after the initial implementation is complete. Each part of the system is built incrementally. Each piece needs to be completed before the next is started. If you don't work this way you can end up with a large amount of work in process. This quickly becomes an unstable system that can jeopardize the project.

What does it mean for a small feature to be done?

- A clear definition of how the feature works is created.
- A test verifies that the feature works properly.
- The code is written to implement the minimum requirements.
- The code has an optimal structure and style (with no duplication).
- The code is committed, tested in the system, and deployed.
- Any open issues that are associated are resolved.

A feature is complete when you never have to think about it again. This requires all of the preceding activities to be completed. If you truly work in these small steps then a simple feature can often be done in fewer than 10 lines of code and thirty minutes of effort. It is possible do a lot of features each day with this method!

When a feature is complete you can think of it as being closed to modification, but open to extension. You shouldn't have to revisit the code for its existing functionality, but you may to improve the structure or functionality when using it for a new purpose.

Flexibility in software comes from extending fully implemented and robust components. Adding functionality is easy if the design and code is simple and clean. By finishing your work before moving on you will have a firm basis for what comes next.

It is impossible to refactor code unless you have unit tests. A typical feature is added by adding one line of code. A unit test can then be created by another line of code. In order to make the structure more readable you will likely add a function definition to the product and also the test code. Add a couple of comments to tell someone else what you are doing.

Consciously count the lines of text you type. Create a set of personal guidelines and a budget for the code you write. Constantly strive to write less code. 100 lines to code a certain function is twice as good as 150 lines, maybe more. Remember complexity is not linear.

Made in the USA
San Bernardino, CA
12 March 2018